About Island Press

Since 1984, the nonprofit organization Island Press has been stimulating, shaping, and communicating ideas that are essential for solving environmental problems worldwide. With more than 1,000 titles in print and some 30 new releases each year, we are the nation's leading publisher on environmental issues. We identify innovative thinkers and emerging trends in the environmental field. We work with world-renowned experts and authors to develop cross-disciplinary solutions to environmental challenges.

Island Press designs and executes educational campaigns in conjunction with our authors to communicate their critical messages in print, in person, and online using the latest technologies, innovative programs, and the media. Our goal is to reach targeted audiences—scientists, policymakers, environmental advocates, urban planners, the media, and concerned citizens— with information that can be used to create the framework for long-term ecological health and human well-being.

Island Press gratefully acknowledges major support of our work by The Agua Fund, The Andrew W. Mellon Foundation, The Bobolink Foundation, The Curtis and Edith Munson Foundation, Forrest C. and Frances H. Lattner Foundation, The JPB Foundation, The Kresge Foundation, The Oram Foundation, Inc., The Overbrook Foundation, The S.D. Bechtel, Jr. Foundation, The Summit Charitable Foundation, Inc., and many other generous supporters.

The opinions expressed in this book are those of the author(s) and do not necessarily reflect the views of our supporters.

Brilliant Green

Brilliant Green

THE SURPRISING HISTORY AND SCIENCE
OF PLANT INTELLIGENCE

Stefano Mancuso and Alessandra Viola

Translated by Joan Benham

Foreword by Michael Pollan

ISLANDPRESS

Washington | Covelo | London

Original title: *Verde brillante: Sensibilità e intelligenza del mondo vegetale*
© 2013 Giunti Editore S.p.A. Firenze-Milano.
www.giunti.it

English edition: © 2015 by Island Press
Translation copyright © 2015 by Joan Benham
Foreword copyright ©2015 by Michael Pollan

The translation of this work has been funded by SEPS
SEGRETARIATO EUROPEO PER LE PUBBLICAZIONI SCIENTIFICHE

Via Val d'Aposa 7
40123 Bologna
Italy
seps@seps.it | www.seps.it

Library of Congress Control Number: 2014956813

Printed on recycled, acid-free paper

Manufactured in the United States of America
10 9 8 7 6 5 4

Keywords: Animal cells, Aristotle, botany, Charles Darwin, colony,
communication, evolution, intelligence, natural selection, network, plant
cells, plant neurobiology, plants, root system, senses, sleep, Venus flytrap

Contents

Foreword

BY MICHAEL POLLAN

Most people who bother to think about plants at all tend to regard them as the mute, immobile furniture of our world—useful enough, and generally attractive, but obviously second-class citizens in the republic of life on Earth. It takes a leap of imagination over the high fence of our self-regard to recognize not only our utter dependence on plants, but also the fact that they are considerably less passive than they appear, and in fact are wily protagonists in the drama of their own lives—and ours.

Brilliant Green will give you a bracing boost over that fence, and put you down in a place where everything—ourselves included—suddenly looks completely different. Chances are, you will come away convinced that it is only human arrogance, and the fact that the lives of plants unfold in what amounts to a much slower dimension of time, that keeps us from appreciating their intelligence—yes, intelligence—and consequent success in the game of life, which has been extraordinary, and dwarfs our own. Plants dominate every terrestrial environment, composing ninety-nine percent of the biomass on Earth. By comparison, humans and all the other animals are, in the words of this intellectually exhilarating book, only a trace.

In Stefano Mancuso, the plants have found their most eloquent or impassioned human advocate since Charles Darwin, who famously wrote that, "It has always pleased me to exalt plants in the scale of organized beings." Stefano Mancuso is a leading researcher—he is a plant physiologist—in the relatively young, and still somewhat controversial, field of "plant intelligence." That term strikes many plant scientists as tendentious, or just over the top, but as soon as you define intelligence as, very simply, the ability to solve the problems that life presents, it becomes impossible to deny such a capability to plants. We jealously guard terms like *intelligence*—and *learning* and *memory* and *communication*—as the monopolies of animals, but *Brilliant Green* makes a persuasive case that these are all qualities we now must share.

I first met Stefano Mancuso in 2013 at his lab, provocatively titled the International Laboratory of Plant Neurobiology at the University of Florence, when I was researching an article for the *New Yorker* on plant intelligence. He shared with me that his conviction that humans fail to perceive the reality of plant life had its origins in a science fiction story he remembers reading as a teenager. A race of aliens living in a radically sped-up dimension of time arrive on Earth and, unable to detect any movement in humans, come to the logical conclusion that we are "inert material" with which they may do so as they please. The aliens proceed ruthlessly to exploit us. (Mancuso subsequently recalled that the story was actually a somewhat mangled recollection of an early *Star Trek* episode called "Wink of an Eye," easy to find online and well worth seeing.)

That formative leap of imagination—opening him up to a plant's-eye view of us, speedy, heedless, and arrogant—has inspired Mancuso's scientific research and now, with science writer Alessandro Viola, this marvelous collaboration. But while *Brilliant Green* is most assuredly *not* a work of science fiction—there is good science to back up its every claim—it is, like the best science, the product of a powerful imagination, one with the ability to see the world from a completely fresh and unencumbered point of view—and to communicate that perspective to the rest of us. So, put aside for a couple of hours your accustomed anthropocentrism, and step into this other, richer, and more wonderful world. You won't regret it, and you won't emerge from it ever quite the same again.

Introduction

Are plants intelligent? Do they solve problems and communicate with their surroundings—with other plants, insects, and higher animals? Or are they passive, unfeeling organisms without a trace of individual or social behavior?

Differing answers to such questions date back to ancient Greece, when philosophers of opposing schools of thought argued for and against the proposition that plants have a "soul." What drove their reasoning? And above all, after centuries of scientific discovery, why is there still disagreement about whether plants are intelligent? Surprisingly, many of the points raised today are the same ones raised centuries ago, and hinge not on science but on sentiment and cultural preconceptions that have existed for thousands of years.

Although casual observation may suggest that the plant world's level of complexity is pretty low, over the centuries the idea that plants are sentient organisms which can communicate, have a social life, and solve problems by using elegant strategies—that they are, in a word, *intelligent*—has occasionally raised its head. Philosophers and scientists in different times and cultural contexts (from Democritus to Plato, from Linnaeus to Darwin, from Fechner to Bose, to mention only some of the best known) have embraced the belief that plants have much more complicated abilities than are commonly observable.

Until the mid-twentieth century there were only brilliant intuitions. But discoveries over the past fifty years have finally shed light on this subject, compelling us to see the plant world with new eyes. In the first chapter we'll explain this, and we'll see that even today, arguments for denying plants' intelligence rely less on scientific data than on cultural prejudices and influences that have persisted for millennia.

The time seems ripe for a change in our thinking. On the basis of decades of experiments, plants are starting to be regarded as beings capable of calculation and choice, learning and memory. A few years ago, Switzerland, amid much less rational polemics, became the first country in the world to affirm the rights of plants with a special declaration.

But what are plants, really, and how did they come to be the way that they are? We humans have lived with them from the time we appeared on Earth, yet we can't say we know them at all. This isn't just a scientific or cultural problem; it goes much deeper. The relationship between humans and plants is so difficult because our evolutionary paths have been so different.

Like all animals, humans are endowed with unique organs, and thus every human being is an indivisible organism. But plants are sessile—they can't move from one place to another—and so they've evolved in a different way, constructing a modular body without individual organs. The reason for such a "solution" is obvious: if an herbivorous predator removed an organ whose function couldn't be performed by another part of the plant, that *ipso facto* would cause the plant's death.

Until now, this basic difference from the animal world has been one of the main obstacles to our understanding and recognizing plants as intelligent beings. In the second chapter, we'll try to explain how this difference occurred. We'll see how every plant has the ability to survive massive predation, and that what ultimately distinguishes a plant from an animal is its divisibility: its being equipped with numerous "command centers" and a network structure not unlike the Internet's. Understanding plants is becoming more and more important. They enabled our coming into existence on Earth (through photosynthesis, creating the oxygen that made animal life possible), and today we still depend on them for our survival (they are at the base of the food chain). They're also the origin of energy sources (fossil fuels) that have sustained our civilization for thousands of years. Thus they are precious "raw materials," essential for our food, medicine, energy, and equipment. And we're growing increasingly dependent on them for our scientific and technological development.

In the third chapter we'll see that plants have all five senses that humans do: sight, hearing, touch, taste, and smell—each developed in a "plant" way, of course, but no less real. So from this point of view could we say they resemble us? Not at all: they're much more sensitive, and besides our five senses, they have at least fifteen others. For example, they sense and calculate gravity, electromagnetic fields, and humidity, and they can analyze numerous chemical gradients.

Though the idea doesn't jibe with our general impression of plants, they may be more like us in the social sphere. In the fourth chapter we'll see how plants use their senses to orient themselves in the world, interacting with other plant organisms, insects, and animals, communicating with each other by means of chemical molecules and exchanging information. Plants talk to each other, recognize their kin, and exhibit various character traits. As in the animal kingdom, in the plant world some are opportunists, some are generous, some are honest, and some are manipulators, rewarding those that help them and punishing those that would do them harm.

Then how can we deny that they are intelligent? The question comes down to terminology, and it depends on how we choose to define *intelligence*. In the fifth chapter we'll see that intelligence can be construed as "problem-solving ability," and that by this definition plants are not just intelligent but brilliant at solving the problems related to their existence. To start with, they don't have a brain like ours, yet they are able to respond adaptively to external stresses and, though using this word about a plant may seem strange, to be "aware" of what they are, and of their surroundings.

It was Charles Darwin who, on the basis of solid, quantifiable scientific data, first suggested that plants were much more advanced organisms than they were thought to be. Today, almost a century and a half later, a compelling body of research shows that higher-order plants really are "intelligent": able to receive signals from their environment, process the information, and devise solutions adaptive to their own survival. What's more, they manifest a kind of "swarm intelligence" that enables them to behave not as an individual but as a multitude—the same behavior seen in an ant colony, a shoal of fish, or a flock of birds.

Plants could live very well without us, in general. But without them we would die out very quickly. And yet in many languages (including our own), expressions such as "to vegetate" or "to be a vegetable" are used to indicate a condition of life reduced to the minimum.

"Vegetable, to whom?" . . . If plants could speak, maybe that would be one of their first questions to us.

CHAPTER 1

The Root of the Problem

In the beginning, there was green: a chaos of plant cells. Then God created the animals, ending with the noblest of them all: man. In the Bible, as in many other cosmogonies, man is the supreme fruit of the divine work, the chosen one. He appears near the end of Creation, when everything awaits him: ready to be subjugated and ruled by the "master of Creation."

In the Biblical account, the divine work is completed in a time frame of seven days. Plants are created on the third day, while the most presumptuous of all living creatures comes into the world—at last—on the sixth. This sequence approximates present-day scientific findings, according to which living cells capable of performing photosynthesis first appeared on the planet more than three and a half billion years ago, while the first Homo sapiens, so-called modern man, only appeared 200,000 years ago (a few seconds ago, in the evolutionary time frame). But arriving last hasn't kept human beings from feeling privileged, even though current knowledge on the subject of evolution has drastically reduced our role of "master of the universe," downgrading our status to that of "newcomer"—a relative position that brings no *a priori* guarantee of supremacy over other species, despite what our cultural conditioning would have us believe.

The idea that plants possess a "brain" or a "soul," and that even the simplest plant organisms can feel and react to external stresses, has been proposed over the centuries by numerous philosophers and scientists. From Democritus to Plato, from Fechner to Darwin (to cite only a few examples), some of the most brilliant minds of all time have been exponents of the intelligence of plants, some attributing to them the capacity to feel, others imagining them as humans with their heads in the ground: sensitive living beings, intelligent and endowed with all human faculties, except those precluded by their . . . odd position.

Dozens of great thinkers have theorized and documented the intelligence of plants. Yet the belief that plants are less intelligent and evolved beings than invertebrates, and that on an "evolutionary scale" (a concept without basis in fact but still fixed in our mentality) they're barely above inanimate objects, persists in human cultures everywhere and manifests itself in our everyday behavior. No matter how many voices are raised in support of recognizing plant intelligence on the basis of experiments and scientific discoveries, infinitely more oppose this hypothesis. It's as if by tacit agreement religions, literature, philosophy, and even modern science promulgate in Western culture the idea that plants are beings endowed with a level of life (not to speak of "intelligence," for the moment) lower than that of other species.

Plants and the Great Monotheistic Religions

"And of every living thing of all flesh, two of every sort shalt thou bring into the ark, to keep them alive with thee; they shall be male and female. Of fowls after their kind, and of cattle after their kind, of every creeping thing of the earth after his kind; two of every sort shall come unto thee, to keep them alive." With these words, according to the Old Testament, God told Noah what to save from the universal Flood so that life would continue on our planet. Obeying God's instructions, before the Flood Noah loaded onto the ark birds, animals, and every creature that moved: "clean" and "unclean" creatures, in pairs, to assure the reproduction of every species.

And plants? Not a word about them. In Holy Scripture the plant world not only isn't considered equal to the animal world, it isn't considered at all. It is left to its fate, probably to either be destroyed by the Flood or to survive it along with other inanimate things. Plants were so unimportant that there was no reason to care about them.

And yet the contradictions this passage contains are soon evident. The first becomes obvious as the narrative continues. After the ark's slow coming aground, when the rain has stopped for several days, Noah sends a dove to bring back news of the world. Is there dry land anywhere? Are there places above water nearby? Are they inhabitable? The dove returns with an olive branch in its beak: a sign that some lands have reemerged and that on them life is possible again. Noah therefore knows (even if he doesn't say it) that without plants there can be no life on Earth.

The dove's news is soon confirmed, and in a short while the ark has come to rest on Mount Ararat. The great patriarch debarks, lets the animals off, and then gives thanks to God. His duties are fulfilled. And what does Noah do next? He plants a vineyard. But where does the original vine come from, if it isn't mentioned elsewhere in the story? Noah brought it with him before the Flood, aware of its usefulness, though not that it was a living being.

In this way, almost without the reader's realizing it, the idea that plants are not living creatures comes through the story in Holy Scripture. In Genesis, two plants, the olive and the grape vine, are associated with the value of rebirth and of life, though the vital quality of the plant world in general goes unrecognized.

All three of the Abrahamic religions have implicitly failed to recognize that plants are living beings, in effect grouping them with inanimate objects. Islamic art, for example, respecting the prohibition against representing Allah or any other living creature, is passionately devoted to the representation of plants and flowers, so much so that the floral style is emblematic. Without stating it outright, this shows the belief that plants are not living beings—otherwise representing them would be forbidden! In the Koran, there is actually no explicit ban on representing animals; the prohibition is transmitted through the *hadith*, the sayings of the prophet Mohammad that form the basis for the interpretation of

Islamic law, by virtue of the fact that in Islam there is no God but Allah and everything comes from him, and everything is him—which evidently doesn't mean plants.

The relationship between humans and plants is totally ambivalent. For example, the same Judaism which is based on the Old Testament forbids the gratuitous destruction of trees and celebrates the new year of trees (Tu Bishvat). The ambivalence comes from the fact that on the one hand we humans are intimately aware that we can't exist without plants, and on the other hand we're unwilling to recognize the role they play on the planet.

It's true that not all religions have the same relationship to the plant world. Native Americans and other indigenous peoples recognize its undeniable sacredness. If some religions have sacralized plants (or rather, parts of them), others have gone so far as to hate or even demonize them. For example, during the Inquisition, plants believed to be used in potions by women accused of witchcraft—garlic, parsley, and fennel—were put on trial along with the witches! Even today, plants with psychotropic effects receive special treatment: some are banned altogether (How do you ban a plant? Could you ban an animal?), others are regulated, still others are considered sacred and used by shamans in tribal ceremonies.

The Plant World According to Writers and Philosophers

Hated, loved, ignored, or sacralized, plants are part of our lives and so of our art, folklore, and literature. In the works they create, the imagination of artists and writers helps construct a vision of the world. What does art tell us about the relationship between human beings and the plant world? Though there certainly are important exceptions, in general, writers depict the plant world as a static, inorganic part of the countryside, passive as a hill or a mountain chain. Consider, for example, *Robinson Crusoe* (1719) by Daniel Defoe, where plants are depicted as part of the landscape but never as living organisms. For the first hundred pages, the whole plot of the novel is based on Robinson's search for other living organisms on the island . . . while he is literally surrounded by them in the form of plants. More recently, in *Suddenly in the Depths of the Forest* (2005) by Amos Oz, a small village is under a curse that prevents any form of life except humans . . . while the village is completely encircled by the plants of the forest.

In philosophy, as we have noted, inquiries into plants' nature have animated the discussions of great minds for centuries. Whether plants had life (or a "soul," as they called it then) was an endlessly debated question centuries before Christ. In Greece, birthplace of Western philosophy, opposing positions on this matter long coexisted: on one side Aristotle of Stagira (384/383–322 BCE) thought that the plant world was closer to the inorganic world than to the world of living things; on the other, Democritus of Abdera (460–360 BCE) and his followers showed a high estimation of plants, even comparing them to human beings.

In classifying living things, Aristotle divided them according to the presence or absence of soul, a concept which for him had nothing to do with spirituality. To understand it, we need to consider the root of the word *animate*, which even today means "having the ability to move." In one of his works, he wrote: "Two characteristic marks have above all others been recognized as distinguishing that which has soul in it from that which has not—movement and sensation" *(On the Soul)*. On the basis of this definition, and with the support of such observations as were possible in those times, Aristotle initially considered plants to be "inanimate." But then he had to reconsider. After all, plants could reproduce! How could one argue that they were inanimate? The philosopher then opted for a different solution and gave them a low-level soul, a plant soul created expressly for them, which in practical terms only permitted them reproduction. If plants couldn't be thought of as equal to inanimate things, because they could reproduce, still—Aristotle decided—they shouldn't be considered all that different from them, either.

Aristotelian thinking influenced Western culture for many centuries, especially in certain disciplines such as botany, where it held sway almost until the beginning of the Enlightenment. So it's little wonder that philosophers long considered plants to be "immobile" and not worth further consideration.

However, from antiquity to the present day, some philosophers have paid the highest honors to the plant world. For example, almost a century before Aristotle, Democritus described plants in a completely different way. His philosophy was based on atomistic mechanics: every object, even if it appeared to be immobile, was composed of atoms in continuous motion, separated in a vacuum. According to this vision of reality, everything moved, and thus at an atomic level even plants were mobile. Democritus even compared trees to upside-down humans, with their head set in the ground and their feet in the air—an image that would often recur through the centuries.

The Aristotelian and Democritean conceptions in ancient Greece thus often gave rise to a kind of unconscious ambivalence, which held that plants were simultaneously inanimate beings and intelligent organisms.

The Fathers of Botany: Linnaeus and Darwin

Carl Nilsson Linnaeus (1707–1778), usually known as Carl Linnaeus, was a physician, explorer, and naturalist whose many interests included the classification of all plants. For this reason, he is often known as "the great classifier," which only partly does him justice, since in addition to his work of classification he conducted intensive research throughout his lifetime.

Linnaeus's ideas concerning the plant world were idiosyncratic almost from the start. First, he identified "reproductive organs" in plants, and he made the "sexual system" the principal taxonomic criterion upon which he based his work of classification. In a bizarre contradiction, this decision earned him both the first university chair and also condemnation for "immorality." (It was known that plants had a sex. But studying this in order to classify plants? . . . how scandalous.) Then the scientist proposed another innovative theory, which only by accident drew less criticism than the first: Linnaeus maintained, with surprising determination and simplicity, that plants . . . sleep.

Even the title of *Somnus Plantarum (The Sleep of Plants)*, his treatise of 1755, didn't observe the caution used by scientists in those days to protect their theories from possible attacks. In fact, based on scientific knowledge of that time and on his own observations of the different positions assumed by the leaves and branches during the night, it was relatively easy for Linnaeus to assert that plants sleep. But it would be several centuries before sleep was recognized as a fundamental biological function related to the brain's most evolved activities, and so his idea was not even contested.

Today the same theory has plenty of opponents, and even Linnaeus, if he had known the many functions of sleep, would probably interpret his own observations differently and would deny the existence of an activity in plants that could be compared to an activity of animals. In fact, he did deny it in another instance: that of insectivorous plants. Linnaeus was quite familiar with plants that ate insects, such as Dionaea muscipula (the Venus flytrap), for example. And he certainly had the experience of observing one as it enclosed, trapped, and digested an insect. Yet that reality (a plant eating an animal) was so incompatible with the rigid pyramidal organization of nature, in which plants were relegated to the lowest level of life, that Linnaeus, like his contemporaries, sought a myriad of other possible explanations rather than acknowledge plain evidence. Without any regard to scientific confirmation of his assertions, from time to time he therefore hypothesized that the insects didn't die at all, and that they chose to remain inside the plant of their own volition and for their own convenience, or that they landed on the plant by chance and not because they were attracted to it. Or even that the plant trap closed by chance, and so couldn't possibly lure an animal. Ambivalence toward the plant world still had its hold on the mind of the great Swedish botanist!

Not until Charles Darwin published his treatise on insectivorous plants in 1875 did a scientist finally assert the existence of plant organisms that feed on animals. But even Darwin, with his characteristic caution, didn't go so far as to call them "carnivores" (as we do today), though he was perfectly aware of plants that prey on rats and other small mammals, such as several supercarnivores belonging to the genus Nepenthes. Some "insectivores!"

We shouldn't be dismayed by Darwin's caution, any more than we are by Galileo's, or the caution of other scientists in centuries past. It's because of their "diplomacy," in fact, that certain revolutionary ideas could slowly filter through the collective consciousness—and into a scientific community that was very conservative. But let's return to Linnaeus for a moment, and ask ourselves: how was it possible for him to assert so boldly that plants sleep, without being shunned or persecuted by his peers? This isn't hard to answer: for a long time it was thought that his theory had no basis in fact, so it wasn't even worth refuting. And furthermore, who cared whether plants slept, when sleep wasn't believed to have any particular function?

Today, we know how many important vital and cerebral functions are linked to this physiological process. But until the turn of this century, even modern science maintained that only the most evolved animals sleep. In 2000, this was disproved by the Italian neuroscientist Giulio Tononi, who showed that even the fruit fly, one of the "simplest" insects in existence, takes its well-deserved rest. Then why shouldn't plants? Maybe the only possible explanation is that this idea doesn't fit with how we think about the vegetal world.

Humans Are the Most Evolved Beings on the Planet. Or Are They?

With few or no exceptions, unfortunately, the idea of the plant world and the so-called Pyramid of Living Things that we've taken with us down through the centuries is the one contained in the *Liber de sapiente* (Book of Wisdom), published in 1509 by Charles de Bovelles (c. 1479–1567). An illuminated illustration from the book is worth more than a thousand words: it shows the living and nonliving species in ascending order. It starts with rocks (which are given the following lapidary comment: *Est*, meaning they exist and that's all; they have no further attributes), continues to plants (*Est et vivit:* thus a plant exists and is alive, but nothing more) and animals (*Sentit:* an animal is endowed with senses), and finally comes to man (*Intelligit:* only man has the faculty of understanding).

The Renaissance idea that among living creatures, some species are more or less evolved and endowed with greater or lesser vital capacities, is still in vogue. It is part of our cultural humus, and nearly impossible for us to give up, despite the passage of more than 150 years since the publication in 1859 of *The Origin of Species*, the foundational work given us by Charles Darwin to understand life on our planet—a book so important that the great biologist Theodosius Dobzhansky wrote: "Nothing in biology makes sense except in the light of

Figure 1-1. Charles de Bovelles's "Pyramid of the Living," from *The Book of Wisdom* (1509). Our way of looking at the natural world hasn't changed much.

evolution." The theories of the great British scientist, who was a biologist, botanist, geologist, and zoologist, are now part of humanity's scientific inheritance. Yet the idea that plants are passive beings, without sensation or any capacity for communication, behavior, or computation—which comes from a completely erroneous view of evolution—is still strongly rooted even in the scientific community.

It was Darwin who proved beyond any reasonable doubt that the question should not be put in those terms, because there are not more- and less-evolved organisms; from the Darwinian point of view, all living beings now populating the earth are at the end of their evolutionary branch—otherwise, they would be extinct. This is a very important assumption since, for Darwin, being at the end of one's evolutionary chain means to have shown, over the course of evolution, extraordinary capacities for adaptation. Of course, the genius naturalist knew well that plants are extremely sophisticated and complex creatures, with many capacities beyond those that are commonly recognized. He devoted a great part of his life and work to botanical studies (some six volumes and about seventy essays), illustrating through them the theory of evolution that brought him imperishable fame. Yet the vast amount of research on the plant world carried out by Darwin has always been treated as secondary: further demonstration—if any were needed—of the scant consideration plants have always received in science.

In his book *One Hundred and One Botanists*, published in 1994, Duane Isely stated: "More has been written about Darwin than any other biologist who ever lived. . . . Curiously, in light of this flood, he is rarely presented as a botanist. True, the fact that he wrote several books about his research on plants is mentioned in much Darwinia, but it is casual, somewhat in the light of 'Well, the great man needs to play now and then.'" Darwin wrote and affirmed several times that he considered plants to be the most extraordinary living things he had ever encountered ("It has always pleased me to exalt plants in the scale of organised beings," he confessed in his autobiography), a theme that he took up again and amplified in his fundamental *The Power of Movement in Plants*, published in 1880. Darwin was a scientist of the old school: he observed nature and deduced its laws. Though not a dogged experimenter, in this book he explains the results of hundreds and hundreds of experiments he carried out with his son Francis, describing and interpreting the innumerable movements of plants: a great many different movements, which involved in most cases not the aerial part but the root, in which he was able to identify a sort of "command center."

For the English naturalist, the last paragraph of his works is always the most important. It is where he presents his final considerations on the subject under discussion, in a way that makes them simple and accessible to everyone. Here is a marvelous example from the famous epilogue to the *Origin of Species*:

> There is grandeur in this view of life, with its several powers, having been originally breathed by the Creator into a few forms or into one; and that, whilst this planet has gone cycling on according to the fixed law of gravity, from so simple a beginning, endless forms most beautiful and most wonderful have been, and are being, evolved.

In the final, expressive paragraph on the movement of plants, the scientist clearly states his conviction that in the roots there is something similar to the brain of a lower animal (an important assertion, to which we will return in chapter 5). In fact, a plant has thousands of root tips, each endowed with its own "computing center," a phrase we use to make plain to even the most spiteful critics that from Darwin on, no one has ever thought or said that in the root there's an actual brain—walnut-shaped and resembling a human's—

which somehow escaped notice for millennia. Instead, the hypothesis is that at a plant's root tip there is a kind of plant analog, endowed with many of the same functions as an animal brain. What's so shocking about that?

Though Darwin's assertions were potentially of great consequence, he was careful not to elaborate on them in his books. Already old when he wrote *The Power of Movement in Plants,* Darwin was certain that plants should be considered intelligent organisms, but he also knew that saying so would stir up a hornet's nest of controversy about his studies. Remember, he'd already had problems defending his theory that humans descended from apes! And so he left the task of developing his thesis to others, especially to his son Francis.

Profoundly influenced by his father's ideas and research, Francis Darwin (1848–1925) carried on Charles's work, becoming one of the first professors of plant physiology in the world and writing the first treatise in English on this new field of study. At the end of the nineteenth century, it still seemed paradoxical to associate the two ideas (plants and physiology). But Francis, who had studied plants and their behavior at his father's side for many years, had become convinced of their intelligence. Now a world-famous scientist in his own right, on September 2, 1908, at the opening of the annual congress of the British Association for the Advancement of Science, he threw caution to the wind and declared that plants are intelligent beings. This provoked the expected storm of protest, but he repeated the assertion, even publishing a thirty-page article in *Science* the same year.

The impact was extraordinary. The debate was reported in newspapers all over the world and divided scientists into

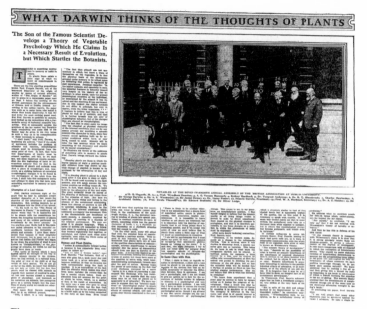

Figure 1-2. *New York Times* page reporting Francis Darwin's announcement at the 1908 annual meeting of the British Association for the Advancement of Science: Plants have a primitive form of intelligence.

two opposing camps. One side—persuaded by the evidence Francis Darwin offered to support his assertions—quickly affirmed the existence of plant intelligence; the other side adamantly rejected the possibility. Just like in ancient Greece!

Years before this debate, Charles Darwin had had a most fruitful correspondence with an Italian botanist in Liguria who is now forgotten—unjustly, since he was one of the most important naturalists of his time and can even be credited with creating the field of plant biology. Federico Delpino (1833–1905), director of the Botanical Garden of Naples, was an outstanding scientist. Through his correspondence with Darwin, he had become convinced of plants' intelligence, and he devoted himself to field experiments studying their faculties, concentrating over a long period of time on so-called myrmecophilia, the symbiosis some plants establish with ants (the term comes from the Greek *murmex*, "ant," and *philos*, "friend"). Charles Darwin was well aware that many plants produce nectar even apart from the flower (though obviously, most of it is produced in the flower in order to attract insects and utilize them as pollen vectors during pollination), and he had also observed that nectar, being very sweet, attracts ants. But he had never studied the phenomenon closely, being convinced that "extrafloral" nectar production (so called because it occurs outside the flower) was essentially due to the elimination of waste substances by the plant. On this point, however, Delpino completely disagreed with the great man. Nectar is a very energy-rich substance that plants produce at great cost to themselves. So, he wondered, why would they get rid of it? There had to be another explanation.

Starting from his observations of ants, Delpino reached the conclusion that myrmecophilic plants secrete nectar in parts of themselves other than their flowers expressly to attract ants and make advantageous use of them as a defensive strategy: the ants, satisfied with their meal, in return for the food, defend the plants from herbivores, like real warriors. Have you ever leaned against a plant or a tree and jumped away from the bites of these feisty little hymenopterans? Ants come to the defense of their host plant instantly, lining up, surrounding the potential predator, and forcing it to retreat! It would be hard to argue that this behavior isn't extremely convenient for both species.

According to entomologists, in fact, ants carry out very intelligent behavior, defending their source of food. For botanists, however, the story has always been (and still is) very different. Not many are willing to say that the plant's behavior also is intelligent (and purposeful) and that the secretion of nectar is a deliberate strategy for acquiring that unusual army of bodyguards.

Plants: Always Second Fiddle

By now, it should come as no surprise that many extraordinary scientific discoveries resulting from experimentation with plants have taken decades to be "confirmed" by research conducted on animals. Discoveries about fundamental mechanisms of life, essentially ignored or greatly undervalued as long as they pertained to the plant world, suddenly become famous when they concern the animal world.

Consider the experiments conducted on peas by Gregor Johann Mendel (1822–1884): they actually marked the beginning of genetics, but for forty years his conclusions were almost completely ignored, until the first genetics boom began, with experiments on animals. Or look at the experience, which had a happy ending for a change, of Barbara McClintock (1902–1992), who won the Nobel Prize in 1983 for her discovery of genome lability. Before McClintock proved otherwise, it was thought that genomes (that is, the entire genetic makeup) were fixed and could not vary over the lifetime of a living being. The "stability of the genome" was untouchable scientific dogma. During the 1940s, with a series of experiments on corn, McClintock discovered that this principle wasn't unassailable at all.

It was a fundamental discovery—so why was it awarded the Nobel Prize forty years later? The reason is simple: her research was carried out on plants, and since McClintock's observations ran counter to academic orthodoxy, she was marginalized by the scientific community for a long time. But in the early 1980s, analogous research carried out on animals confirmed the existence of genome lability in other species, and this "rediscovery," not only her own research, won McClintock the Nobel Prize and recognition of her own contribution.

Of course, genome lability is far from the only example of such discoveries. There's a pretty long list, from the discovery of the cell (which was first made in plants) to RNA interference, which won Andrew Fire and Craig C. Mello the 2006 Nobel Prize. That was essentially a "rediscovery" on a worm *(Caenorhabditis elegans)* of findings made by Richard Jorgensen on petunias twenty years earlier. And the upshot? Nobody knows about the research on petunias, while the research on a very lowly worm (but an animal) merited a Nobel Prize for Physiology and Medicine.

There are many more examples, but the basic story is the same: the plant world always gets second ranking, even in academia. Yet plants are often used in research because of the similarity between their physiology and that of animals, not to mention that experimentation on these organisms raises fewer ethical problems. But are we really sure that the ethical implications are inconsequential? We hope that reading this book will help plant some doubts on that score.

When the absurd subjection of the plant world to the animal world finally comes to a halt, it will be possible to study plants—much more usefully—for their differences from animals, rather than their similarities to them. New and fascinating frontiers for research will open up. But we might be forgiven for asking: What brilliant researcher would devote herself to plants instead of animals, knowing that she will be excluded from the majority of scientific awards?

As we have seen, this state of affairs is a natural outcome of our culture. In life as in science, the common scale of values relegates plants to last place among living things. An entire realm, the plant world, is underappreciated, despite the fact that our survival on the planet and our future depend on it.

CHAPTER 2
The Plant: A Stranger

Human beings have lived with plants since our appearance on the earth about 200,000 years ago. Two hundred thousand years would seem to be enough time to get to know someone. But it hasn't been enough time for us to get to know plants. We know very little about the plant world, and we probably see plants in much the same way as the first *Homo sapiens* did.

This assertion, though patently indemonstrable, may be clarified with a simple example. Let's consider an animal—say, a cat—and try to describe its characteristics. What can we say about the cat? It's smart, clever, affectionate, sociable, opportunistic, agile, quick, and who knows how many other things. Now let's consider a plant—say, an oak tree—and describe its characteristics, too. What is there to say about an oak tree? It's tall, shady, knotty, fragrant . . . what else? At most, we could add some aesthetic qualities and appreciations of its

usefulness. We certainly wouldn't include attributes referring to its "social dimension," whereas in the case of the cat
we've said that it is sociable (though "individualistic" would
also describe a cat's way of relating to its environment). We
wouldn't attribute any sort of intelligence to a plant—whereas
in the cat we recognize it easily—nor would it occur to us to
call an oak affectionate!

And yet something's off about this. How is it possible that
living beings that are unintelligent, without social aptitudes,
and incapable of relating to their environment, have survived
and evolved on the planet? If plants really functioned so
poorly, natural selection would have swept them away long
ago!

But we don't need to look to the past as evidence; in the
last several decades, science has been showing that plants are
endowed with feeling, weave complex social relations, and can
communicate with themselves and with animals, all of which
we will explore in the following chapters. So why do human
beings still see the plant world only as raw material, or a food
source, or decor? What prevents us from going beyond this
initial, superficial valuation of the life forms that populate it?

Euglena versus Paramecium: An Even Match?

Besides the cultural factors we saw in the first chapter, two
others influence our perception of the plant world: an evolutionary factor and a temporal one.

Let's start with the evolutionary factor and attempt an
analysis, first by asking what we mean by the word *evolution*.
Evolution refers to the slow, continuous process of adaptation

to the environment, in the course of which a living organism develops the characteristics most suited to its survival. During this process, each species acquires or loses characteristics and capacities in relation to the kind of habitat in which it lives. Of course, all of this happens over very long periods of time, but it can lead to macroscopic changes between the original and the eventual organism. Evolution has played a fundamental role in differentiating animals from plants, and today it's part of the problem that keeps us from deeply knowing the plant kingdom.

To see this more clearly, let's take a step back.

We know that the first single-celled organisms that appeared on the planet were algae—that is, the plant kind of living things. Through photosynthesis, they created the oxygen that enabled life to spread over the earth. This included the emergence of eukaryotes, or animal cells.

In those days, as today, plant and animal cells were not as different as one might think. To be sure, plant cells are more complex, because compared to animal cells they have an additional organelle—the chloroplast—in which photosynthesis takes place; and a cell wall that surrounds the entire cell, making it far more robust than an animal cell. But these two differences aside, plant and animal cells are really very similar.

So how to explain the fact that when a single-celled plant organism is compared to a single-celled "animal" (so to speak), the latter is always considered more complex, more evolved—in a word, superior?

Let's compare two unicellular beings, one animal and one plant: the paramecium and the euglena. We're taking some

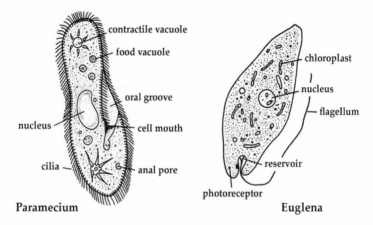

Figure 2-1. Structural comparison of the paramecium and the euglena. The two organisms are very similar, but the second has a primitive eye (a photoreceptor) with which it perceives light.

license in calling the paramecium an animal since, along with other protozoa, it's now in a separate classification, the protists. But until a few years ago, for all intents and purposes, it was considered an animal: as the name *protozoa* implies (from the Greek *protos*, "first," and *zoon*, "animal"), it's a proto-animal.

The paramecium is a minuscule unicellular organism whose body is covered with cilia that act like oars, allowing it to swim and move around in the water. If you look at it under a microscope, you can't help being fascinated by its elegant evolution, and by movements that seem to imply elegant behavior. It's a true champion among living beings: a single cell, but capable of astonishing activity. Writing about another little amoeba-like unicellular animal, Herbert Spencer Jennings (1868–1947) in his book *Behavior of the Lower Organisms*, published in 1906, wondered whether we would be more apt to grant intelligence to the predatory amoeba if it were the size of a whale, and a potential threat to humans.

And in the other corner, we have another marvel of creation, a minuscule single-celled green alga: the euglena. It, too, can be classified with the protists, but without a doubt it has a plant nature.

Looking at such simple living organisms and discovering their extraordinary abilities can help us see what underlies our prejudiced view of the plant world. What do these unicellular organisms have in common, and how are they different from each other? Do the animals really have a minimal form of intelligence, but not the plants?

To get a general idea, let's start with the paramecium. For such a small organism, it has surprising abilities: for example, it can locate food and move to reach it.

Naturally, in order to live, the euglena needs energy, too. Normally, it supplies its energy needs through photosynthesis, like all plants, but if light is scarce, it doesn't give up: it transforms itself into a predator and behaves like an animal. It can locate food and move to reach it—yes, it's a plant, but it moves! This microscopic alga, in fact, swims with the aid of very thin flagella.

Obviously, both the paramecium and the euglena can reproduce. If you watch them moving in the water, there don't seem to be many differences between them. But wait: there are electrical signals traversing the body of the paramecium, transmitting information across a single cell. For this reason, it's been called a "swimming neuron," which seems like a pretty good definition of a paramecium. But there are the same kind of electrical impulses going across the single-celled body of the euglena. So they're even again.

Can the paramecium and the euglena do the same things? Does the match between plants and animals end in a draw? No way—but the outcome isn't what we would expect. The one with the ace up its sleeve isn't the paramecium but the euglena, which has another capacity that beats the competition hands-down: it can carry out photosynthesis. To improve this capacity, it has developed a rudimentary sense of sight, which allows it to intercept light frequencies and then to find the best position for receiving light.

But if the euglena can do everything a paramecium can do, and in addition, can see and produce energy by transforming light from the sun, why has no one ever called it a "swimming neuron" or some other epithet that expresses its exceptional abilities? Hard to say. There's no rational way to explain the general disregard for solid scientific evidence that plant cells have greater capacities than those of animal cells.

Five Hundred Million Years Ago

Returning to the evolutionary obstacle we mentioned at the beginning of the chapter, let's go back about 500 million years, when the differentiation between plants and animals began. The first organisms chose two divergent paths, which can be summed up thus: plants opted for a stationary lifestyle, animals for a nomadic one. It's interesting to note in passing that the same choice in favor of a stationary life gave birth to the first great civilizations.

Plants faced the necessity of obtaining from the earth, the air, and the sun everything they needed in order to live; animals, on the other hand, needed to feed on other animals or on plants, and for that purpose developed manifold movement capacities (running, flying, swimming, and so on). On this account plants are defined as "autotrophic" (from the Greek *autos*, "by itself," and *trophe*, "food"), that is, self-sufficient, not dependent on other living things for their survival; and animals as "heterotrophic" (from the Greek *eteros*, "other," and *trophe*, "food"), because they're not self-sufficient.

From generation to generation, this initial choice led to other fundamental differences between the animal and plant worlds, to the point that now they can be considered the yin and yang, the black and white, of ecosystems. Plants are stationary and animals are mobile; animals are aggressive, plants passive; animals are swift, plants slow. We could come up with dozens of such antithetical pairings, but they would amount to the same thing: life has evolved very differently in the plant world and the animal world over the past 500 million years.

The primitive choice to evolve as beings that are stationary or in motion led, over time, to an extraordinary differentiation in bodies and ways of life: animals have chosen to defend and feed themselves, and to reproduce, through motion (or flight), while plants have chosen to remain fixed in one place, which has imposed on them the necessity of finding solutions that are completely original, at least from our point of view (which, let's not forget, is an animal one).

A Plant Is a Colony

To start with, being stationary and therefore subject to being preyed on by animals, plants developed a kind of "passive resistance" to external attack. Their bodies are constructed on a modular design, in which each part is important but none is truly indispensable. This structure represents a fundamental advantage vis-à-vis the animal kingdom, especially considering the number of herbivores on the planet and the impossibility of escaping their voracious appetites. The first advantage of having a modular organization, to give just one example, is that, for a plant, being eaten isn't that big a deal! Could any animal say that?

The physiology of plants, as we will see, is based on different principles from that of animals. While animals have evolved to concentrate almost all their most important vital functions in a few organs such as the brain, lungs, stomach, and so on, plants have taken into account the reality of being easy prey, and avoided concentrating their faculties in a few neurological areas. It's a bit like not keeping all your money in one place, but instead dividing it up and hiding it in several places to minimize the loss in case of theft, or diversifying your investments to distribute risk. In short, a very wise move!

A plant's functions are not related to organs—which means plants breathe without having lungs, nourish themselves without having a mouth or stomach, stand erect without having a skeleton, and as we will soon see, make decisions without having a brain.

It's because of this very special physiology that large portions of a plant can be removed without putting its survival at risk: some plants can have up to 90 or 95 percent of themselves eaten, but then grow back normally from the small surviving nub. A meadow grazed on by an entire herd can grow back in a few days. You don't have to be an herbivore to experience this phenomenon; if you've ever tried to cut back an ivy or a windweed, or even to clip your lawn, you know what we're talking about. So as an evolutionary strategy, plants, being stationary (or more properly speaking, sessile) organisms, have chosen to be composed of divisible parts in order to better withstand predators. Animals, on the other hand, which based their defensive strategies on movement from the outset, never developed regenerative capacities, or did so only in a few cases. Yes, a lizard can grow back its tail, but a foot, an arm, or its head, once cut off, doesn't grow back. But if part of a plant is removed, it generally not only survives, but sometimes even benefits: consider the rejuvenating effects of pruning. This characteristic is a direct result of its structure, which is very different from ours. A plant is made up of repeating modules: the branches, stem, leaves, and roots are all combinations of very simple modules, which essentially add on to each other independently, a little like Lego blocks.

True, a geranium on a terrace doesn't give that impression: it looks like a unique being. But if you take off a piece and then replant it—if you take a cutting, in gardener's parlance—the piece of geranium will put out new roots and grow into a new plant, whereas neither our arm nor an elephant's foot can regenerate a whole new organism or stay alive apart from the rest of the body.

It's no accident that we continually refer to ourselves as *individuals*: the term comes from the Latin *in* (which here means "not") and *dividuus* ("divisible"). Our body really is indivisible: if we're cut in half, the two halves can't live separately; they die. But if we cut a plant in half, the two parts can still live independently, for the simple reason that a plant isn't an individual. In fact, the right way to think about a tree, a cactus, or a shrub is not to compare it to a human being or any other animal, but to picture it as a colony. A tree is much more like a colony of bees or ants than an individual animal.

Though plants are very ancient, from this standpoint they also turn out to be exceptionally modern. One of the cardinal concepts underlying many of the technologies made possible by the advent of the Internet and based on the interconnection of groups (such as social networks) is that of so-called emergent properties, typical of superorganisms or swarm intelligences. These are properties that single entities develop only by virtue of the unitary functioning of the group; none of the individual components possesses them on their own— just as bees or ants, by forming colonies, develop a collective intelligence much greater than that of their individual members. We'll discuss plant behavior at greater length in chapter 5, on plant intelligence.

A Problem of Tempos

Let's return to the reasons that prevent us from recognizing plants for what they are—social organisms, sophisticated and highly evolved like us. There's another aspect to our inability to perceive the complex reality, one that has to do with time.

We all know that the average lifespan of living creatures varies considerably from species to species: a human being lives about 80 years; a bee less than two months; a giant tortoise more than 100 years. Beyond the variation in average life span, animals also have different vital rhythms: some hibernate; some move and reproduce much faster than we do, others much more slowly. It wouldn't seem to be that hard to recognize the existence of time scales very different from our own. But that isn't the case. A flow of events that gives rise to a time scale so slow as to be imperceptible to our eyes doesn't compute. While these adjectives are obviously meaningless in absolute terms, another way of putting it is that we are "fast" and plants are "slow." Very slow.

The difference in speed between us and them is so great that our perception can't grasp it. It's a bit like a *trompe l'oeil* or an optical illusion, but on a temporal scale. For example, we know very well that a plant moves to capture light, to distance itself from danger, and to seek support (in the case of the climbing plants). For decades, modern techniques of photography and film have enabled us to reconstruct plant movement, which Darwin had already discussed and validated. Today a quick Internet search will bring you to a video showing a flower opening or a shoot growing. Yet in our perception, plants remain "still."

The sight of these movies astounds us, it speaks to the existence of movements in plants, but it doesn't budge our unshakable conviction, partly instinctive, that these creatures are closer to the mineral world than to animal life. Our senses don't perceive plants moving, so we act as if they are inanimate objects. It makes no difference that we know they grow and therefore move; to us they're motionless because their movements escape our sight, and thus our deep understanding.

But what's the significance of our denial? In the hypertechnological society we live in, there are many things of which we have no direct (sensory) knowledge, but whose properties we don't doubt. Few people know how a television works, or a phone or a computer, but we wouldn't think of belittling their technical characteristics merely because we have no direct sensory experience of the ways they work. Our knowledge of the structure of the universe and the composition of matter is mediated by extremely complicated instruments. But who would think of denying the complexity of atomic structure, even though it's much more removed from our sense perception than the structure of plants? Of course, education plays an important role in this regard.

So why doesn't something similar happen with respect to plants? It doesn't seem improbable that there may be a kind of "psychological block" preventing any kind of cultural mediation that over time might mitigate this instinctive behavior on our part. We'll explain what we mean.

Our relationship with plants is one of absolute, primordial dependence, and in that sense it somewhat recalls the relationship of a child to its parents. While we're growing up, and especially in adolescence, we go through a period of totally denying our dependence on our parental figures that frees us to attain psychological autonomy, in preparation for actu autonomy, which will come many years later. It's not out of the question that a similar psychological mechanism enters into our relationships with plants. No one likes depending on another. Dependence coincides with a position of weakness and vulnerability that we don't enjoy contemplating.

We may resent those we depend on, because they don't make us feel completely free. In short, we're so dependent on plants that we do everything we can not to think about them. Perhaps we don't wish to remember that our very survival is linked to the plant world, because that makes us feel weak—hardly masters of the universe! Of course, this argument is partly intended to be provocative, but it may be useful in clarifying the balance of power between us and the plant world.

Life Without Plants: Impossible

If plants disappeared from the earth tomorrow, human life would continue for no more than a few weeks, or at most months. Very soon, all higher forms of life would disappear from the planet. On the other hand, if we were the ones that disappeared, in a few years plants would repossess all the land previously taken over from the natural realm, and in a little less than a century every sign of our enduring civilization would be covered in green. Maybe this can help us take the measure of the relative importance, in biological terms, of plants and human beings.

To use another metaphor: we could say that in biology we're still in a period which we could define as Aristotelian-Ptolomaic. Before the Copernican Revolution, people believed that Earth was at the center of the universe and all the celestial bodies revolved around it—a totally anthropocentric vision, which Galileo endeavored to subvert, and which took centuries to disappear from popular opinion. Well, we could say that biology finds itself in a more or less pre-Copernican situation. The reigning idea is that humans are the most important living beings and everything revolves around us: because we have imposed ourselves on the others, we're the absolute lords of nature. An intriguing and consoling vision . . . if only it were true!

In fact, our situation really isn't so stellar. The plant world alone represents more than 99.5 percent of the biomass of the planet. It's like saying that if 100 is the total weight of everything alive, according to various estimates, between 99.5 and 99.9 percent is composed of plants. Or to put it the other way around, of all living things, animals—humans included—represent only a trace (a scant 0.1 to 0.5 percent).

Despite determined human efforts at maximizing deforestation, plants are the incontestable queens of living things. And thank heaven! It is this relationship that makes life on Earth still possible.

As we know, plants are at the base of the food chain: everything we eat (meat and fish included) either is a plant or became what it is by feeding on plants. It might seem that humans have recourse to the full variety of plants for our nutrition, but this isn't the case. There are principally six plants from which we get most of our calories: sugar cane, corn, rice, wheat, potato, and soy. These and a few others form the nutritional base for almost all humans throughout the world. They are the so-called food plants, very special living beings.

Cultivating a plant is a little like raising an animal. Have you ever wondered why humans' carnivorous diet is based almost entirely on beef, chicken, and pork? Why hasn't any culture based its food on lion, gnu, wolf, bear, or snake? These animals are perfectly edible, on a par with cows and chickens. The answer, obviously, is that domesticated animals are easier to raise. A bear is excellent to eat but not easy to raise. Likewise, it's plain that not all plants lend themselves to intensive cultivation.

Comestible (edible) plants are numerous but most cannot be cultivated on an industrial scale because of the way they've evolved. They're wild, like tigers and bears. Dogs, on the other hand, evolved from wolves as a new species, through discovering that living symbiotically with humans was easier and more convenient than fighting for survival. Over the course of evolution an excellent, mutually satisfying collaboration came about: humans feed and take care of dogs and receive protection and companionship in return. Certain plants have adopted a similar evolutionary strategy: by feeding humans they have been protected from insects, provided with nourishment, and, especially, propagated until they've spread to the far reaches of the planet.

Alimentation is only the first and most intuitive link in our dependence on plants. Then, obviously, there's oxygen. We all know that the oxygen we breathe comes from plants and that our survival depends on the existence of oxygen in the air. But not everyone knows that a large part of the energy we use is also of plant origin, and that we have plants to thank for the energy which has been at our disposal for thousands of years.

Think about it: large quantities of the available energy sources on Earth were once concentrated inside plants, through the transformation of solar energy into chemical energy. This miraculous process, which we call photosynthesis, converts light rays and carbon dioxide present in the atmosphere into sugars, that is, molecules with a high energy content (as anyone on a low-calorie diet who's had to give them up knows). This is the first, fundamental phase in which, through successive transformations, the energy we consume is created, in one of the numerous forms it takes (from wood to charcoal, from petroleum to other combustible fuels).

At the turn of the last century, the Russian botanist Kliment Timiryazev (1843–1920) wrote, "The plant is the connecting link between the earth and the sun," and in fact, almost everything humans use for energy has always come from plants.

Practically speaking, fossil fuels (charcoal, hydrocarbons, oil, gas, and so on) are nothing but underground accumulations of energy from the sun which, during various geologic periods, plant organisms have put into the biosphere by means of photosynthesis. Far from being "minerals," as some persist in calling them, they're true organic deposits.

So after air and food, plants enrich us with another fundamental element: energy. This should be enough to make us worship everything green. And that's even without considering medicine. Practically our entire pharmacopoeia is obtained from molecules that are either produced by plants or synthesized by human beings copying plant chemistry.

Throughout the world, in Eastern and Western cultures, in advanced and developing countries, plants are a fundamental and indispensable component of medicine. Their benefit to human beings comes not only from the pharmacological use of many of the molecules they produce but also directly, in many different environmental contexts, through the positive effects of the presence of plants on our psychophysical well-being.

The benefits plants bring by producing oxygen, absorbing carbon dioxide and pollutants, and moderating climate have been known for a long time. But other ways plants can affect our well-being have been studied only recently, and the findings are remarkable: the presence of plants has been reported to reduce stress, increase attention, and speed recovery from illness.

Simply seeing a plant induces calm and relaxation, as can be shown by measuring physiological parameters. Hospital patients with windows looking out on plants have less need for painkillers and are discharged sooner than patients with views of buildings or empty lots. This is why (for essentially economic reasons) many new hospitals built in northern Europe have spaces (sometimes an entire floor) devoted to plants, where patients can spend their time. In the United States, Boston Children's Hospital and the University of Maryland Rehabilitation and Orthopaedic Institute are two examples among many with gardens open to patients and other visitors.

The effects of the presence of plants on babies and children have been studied recently from a number of different perspectives. And the results of the first studies are striking, to say the least.

One study at the University of Illinois at Urbana–Champaign, for example, looked at student performance on tests which they were asked to do in their rooms. The tests, which required some concentration, had clearly better outcomes for students whose windows looked out on green areas than for students with views of buildings.

Even more than university students, elementary school children show improved attention capacity in the presence of plants, as studies at a school in Florence, Italy, have demonstrated. What's more, there are fewer accidents on tree-lined streets, and there are fewer suicides and fewer violent crimes in neighborhoods with plenty of green spaces. In short, plants positively influence our mood, concentration, learning, and general well-being. The presence of plants seems indispensable even on long space missions, not just as food but for their relaxing effect.

The reasons why plants have these psychological benefits for us are still mostly unknown and may go back far in time, bound up in our unconscious awareness that without them life for our species wouldn't be possible. The calm that pervades us in their presence may be the echo of an ancestral awareness that everything we need and every chance for our survival dwells in the green world. Now as long ago.

CHAPTER 3
The Senses of Plants

It's plain to see that plants don't have eyes, a nose, or ears. So how is it conceivable that they have sight, smell, hearing, and even taste and touch? Everything tells us otherwise: our culture, our senses, and what we gather from simple observation.

We're taught to think that plants "vegetate." In other words, they don't move, they perform photosynthesis, they put out new shoots every so often, sometimes they bloom or lose a leaf, and that's about it.

In our language, the word *vegetable*, when not referring to a plant, has acquired an offensive connotation: for us, "to be a vegetable" or "to be reduced to a vegetable" means to have lost the entire endowment of sensory and motor faculties that was ours even before birth—in effect, to be merely living. Just like plants. Or is it?

As we saw in the first chapter, the idea that the plant world is composed of sensorily deprived beings came to us whole from ancient Greece. It passed unchanged through the Renaissance—as depicted in the famous "Pyramid of Living Things," in which plants exist but neither feel nor think—as well as through the more exacting screening of the Enlightenment and the Scientific Revolution, which in theory should have exposed this model as absurd.

Yet imagine being "reduced" to immobility, or better yet, having chosen it as a useful evolutionary strategy, which already—as we have seen—is the case with plants. Wouldn't it be even more important, then, for you to see, smell, hear, and generally explore the environment sensorially, since you couldn't do so by moving around? The senses are indispensable instruments for life, reproduction, growth, and defense, which is why the plant world never dreamed of doing without them!

As we will see, plants have all five senses, just like us. And that's not all: they have fifteen others. Clearly, these developed according to plant, not human, nature, but that doesn't make them any less reliable.

Sight

Do plants see us? And if so, how? To answer these questions we need to define what we mean by "sight." Obviously, plants don't have eyes. But does that mean they can't see?

Let's go through some dictionaries and eliminate definitions that refer to eyes and then see what we have left. Sight is defined as "the faculty of seeing, of perceiving visual stimuli by means of organs adapted to that function," "the sense that enables perception of visual stimuli," and "the faculty or sense of seeing. The sense of light and of illuminated objects."[1]

Well, then: plants certainly lack eyes, and therefore sight as it's classically understood; but if what we're talking about is a "sense of light" or of "perceiving visual stimuli," the story is completely different. By those definitions, plants not only fully possess a sense of sight, they've developed it in a remarkable way. Plants can intercept light, use it, and recognize its quantity and quality. They've developed this ability for the obvious reason that they get most of their energy from light, through photosynthesis.

The quest for light dominates a plant's life and behavior: having plenty of light, for a plant, is what being rich is for a human. And the opposite is equally true: being in the shade means being poor. In the plant world, as in ours, the greatest daily energy expenditure goes toward sustaining oneself. For plants, this means a constant striving to get and make use of light.

Later, we will see that this wealth or poverty of means affects a plant's development, behavior, faculties, and learning potential, just as it does for humans.

Anyone who has observed a plant, indoors or out, has noticed that it changes position, growing in the direction of light and moving its leaves to maximize the benefit to itself. (This rapid movement is known as "phototropism,"

---

1. These definitions for sight can be found in *Il Sabatini Coletti, Il Grande Dizionario Hoepli*, and the online etymological dictionary Etimo.it, respectively.

from the Greek *phos*, "light," and *trepestai*, "to move"). This only makes sense: for a plant organism, intercepting adequate light is a problem that has to be solved in the quickest and most efficient way possible. So two plants next to each other—for example, in the woods or in a pot—may end up in competition, because the taller plant's leaves put the smaller plant in the shade. The dynamic that causes plants to grow faster to try to surpass a rival is called "escape from shade." It's a strange name: escape isn't a type of behavior that we normally associate with the plant world, while what is triggered here is a real struggle for the conquest of light.

The "escape from shade" phenomenon can be seen so well with the naked eye that it was already perfectly familiar in ancient Greece. Yet, though this typical plant behavior has been recognized for thousands of years, its essential significance continues to be ignored or underestimated. What is it we are actually talking about, after all? Nothing less than a genuine expression of intelligence, which means calculating

Figure 3-1. Example of phototropism, growth of a plant toward the source of light.

risk and estimating benefits: this reality should have been obvious for centuries, if prejudice hadn't clouded our eyes.

Think about it: during its escape, the plant starts growing faster in order to surpass its rival in height and thus obtain more light. But this rapid, intense growth has a very high energy cost to the plant, so high that if the effort doesn't succeed, it may prove fatal. The plant invests energy and materials in a rather expensive and uncertain operation—like an entrepreneur investing for the future. The plant's behavior shows that it can plan and use resources to bring about future results: in short, this is typical intelligent behavior.

But returning to the senses: how does the plant perceive light? Inside the plant, a series of chemical molecules act as photoreceptors that receive and transmit information about the direction from which light rays originate, and about their quality. The plant not only distinguishes light from shade, it recognizes the quality of the light by the length of its waves. Different types of exotically named photoreceptors—phytochromes, cryptochromes, and phototropins—absorb specific wavelengths in the red, far red, blue, and ultraviolet bands. These are the most important wavelengths for the plant since they regulate many aspects of its development, from germination to growth to flowering.

But where are the light receptors? Humans' eyes are in the front of the head: a strategic position from an evolutionary standpoint because they're high up (for better view and wider visual field), near the brain (our one and only) and protected from external attack (we give the head the most protection, since that's where four of our senses and our brain are concentrated). In plants, as we know, things work differently. Plant organisms have evolved so as to avoid concentrating their functions in a single area of the body, and thus circumvent the risk that being a snack for an herbivore will end tragically for the plant.

In the plant world, all faculties are present almost everywhere and no part is truly indispensable. Because of this general structure, even a plant's light receptors are dispensable in great quantities. They're found mostly in the leaves, the organs specialized for the process of photosynthesis, but elsewhere, too. Even the youngest part of the stem, the tendrils, shoots, shoot tips, as well as the wood (what we generally define as "green," which does not easily burn), are rich in photoreceptors. It's as if the whole plant were covered with tiny eyes. The roots are also incredibly light-sensitive; but in contrast to the leaves, they don't like light at all. The leaves tend to grow toward and to face the source of light, showing that they prize it and exhibiting what's called "positive phototropism"; but the roots do exactly the opposite, as if they suffered from a sort of "photophobia" ("fear of light," from the Greek *phos*, "light," and *phobia*, "fear"), which makes them flee from any light source, a behavior known as "negative phototropism."

Here, a practice bears mention that shows once again how lack of knowledge of the plant world can lead to distorted experimental results. It's safe to say everyone knows roots grow in the soil, and so in the dark . . . right? No, in modern laboratories where plant studies are carried out, it seems they never got the news. In molecular biology (the name of the new scientific discipline that has gradually supplanted the glorious terms *botany* and *plant physiology*), experiments are almost always done using model plant seedlings (the most famous species being *Arabidopsis thaliana*, a true modern-day lab star) cultivated not in soil, but on a base of gel or some other transparent support medium containing all the nutrients necessary for normal growth. These substrates facilitate study of the seedlings' behavior, both by their transparency and by making it possible to select the nutrients the plant receives. And their contribution to research really has been very important, except for the small problem just mentioned. In these experiments, the roots are almost always kept in full light—a situation that is completely unnatural and that stresses the entire plant. Roots cultivated on these gels tend to grow very fast and to move a great deal, in an attempt (irretrievably doomed to fail) to escape from the light source that disturbs them. And yet their rapid growth is generally attributed to the health of the plant, which, the thinking goes, grows more roots because it is thriving. On the contrary, the root grows faster because it's trying to get away: a little common sense would tell us that a plant's roots should remain in darkness and not in full light like the leaves.

But it isn't just the roots that seek the dark. There is a precise moment of the year when even the aerial part of certain plants "closes its eyes": autumn, the time when many trees, those known as deciduous, lose their leaves. And if most of a plant's light receptors are concentrated in the leaves, the organs specialized for photosynthesis, what happens when a tree loses them? Exactly what happens to an animal when it closes its eyes: it rests.

Deciduous plants are typical of climates characterized by a relatively cold winter. In tropical and subtropical regions, where the mild climate and constant sunlight promote continuous activity, there are no deciduous plants; instead there are evergreens. But in regions with a temperate or continental climate, the alternation of hot summers and cold winters influences plants' behavior just as it does that of animals. We all know that where winters are very severe, some animals hibernate to survive the cold and the scarcity of food; sleeping is a very efficient way to get through the difficult wintertime. It's so efficient that the plant world has adopted the same strategy. At the first cold, deciduous plants lose their leaves, the part of themselves most delicate and exposed to cold, and at continual risk of freezing during the winter, and go into hibernation. In the plant world, this periodic sleep that protects the organism from a difficult climatic situation is called "vegetative rest," but it's exactly the same concept as hibernation in the animal kingdom. The plant slows down its own vegetative cycle, "closes its eyes," and sleeps through the winter, then resumes functioning normally in the spring, with the formation of buds and new leaves that "reopen their eyes."

While we're still on the subject of sight and eyes, we must finally mention Gottlieb Haberlandt (1854–1945), whose theories baffled the scientific community in the middle of the last century. The great Austrian botanist formed the hypothesis, which he was unable to test experimentally, that plants' epidermal cells function like real lenses that give the plant a fairly clear idea not only of light but also of shapes. Haberlandt hypothesized that plants utilize their epidermal cells just as we use our cornea and lens, to reconstruct real images of the external environment.

Smell

Gottlieb Haberlandt's intriguing theories remain unproven experimentally, so we may still doubt that plants—which definitely are light-sensitive and endowed with the sense of sight—can actually distinguish the shapes of objects. But when it comes to plants' sense of smell, odd as it may seem, we must admit that they really do have superfine "noses." Naturally, we're not talking about sensory organs like our own: plants' sensitivity is not concentrated in one area; and while we smell with our noses only, they smell with their entire body.

To perceive a smell, we breathe in air with our nose and pass it through the olfactory canal, which is lined with chemical receptors that capture the molecules present in the air and produce corresponding nerve signals that carry the smell/information to the brain. Plants' sensitivity to smells is diffuse: imagine our having not one nose but millions of tiny noses scattered all over our body. From the roots to the leaves, a plant is composed of billions of cells, whose surfaces often have receptors for volatile substances able to set off a chain of signals communicating information to the entire organism. Picture these receptors as so many different locks arranged on the surface of the cells, and the smells as so many keys: each lock opens when it comes in contact with the right key, and this sets off the mechanism that produces the olfactory information.

But what is smell for, in the plant world? Plants use "smells"—that is, molecules called BVOCs (biogenic volatile organic compounds)—to receive information from their environment and to communicate with each other and with insects, something that goes on continually (see "Communication Between Plants and Animals" in chapter 4). All smells produced by plants—for example, those of rosemary, basil, or licorice—are equivalent to precise messages: they are the plants' "words," their lexicon! Millions of different chemical compounds function like signs in a real plant language, about which we know very little. We do know that each compound transmits precise information, such as warnings of imminent danger, or messages of attraction or repulsion, or something else. Of course we've always known that every angiosperm (the term for flowering plants, from the Greek *angeion*, "envelope," and *sperma*, "seed") produces a specific odor to communicate with the insects that pollinate it; in this case the message is "private," not meant for other plants, and has one clear purpose. But why do sage, rosemary, or licorice give off their typical scents even when they're not in flower? All we know is that they have their reasons: producing those smells costs energy, and no plant would waste energy uselessly! It's still a long road from this simple observation to the possibility of interpreting these plant messages with certainty.

We could draw a parallel between our present situation and that of the Egyptologist Jean-François Champollion before 1822, the year in which he finally succeeded in decoding Egyptian hieroglyphics: we've figured out that certain signs (smells) correspond to certain messages, but that's a small number out of all the volatile molecules plants give off. And what makes the decoding more difficult is the known fact that the message isn't always associated with a single volatile molecule, but rather with a set of many molecules, each in a fixed proportion to the others. In short, even in plants' language there seems to be a kind of polyphony in keeping with their essential non-individuality: not one voice, but a plurality of accents that make them all the more appealing and interesting.

One day, we will likely find the key for decoding this language. Until then, we'll have to content ourselves with the little we know, and with the meanings we can associate with certain volatile molecules. For example, we know the meaning of "methyl jasmonate," a molecule that many plants produce under stressful conditions. Methyl jasmonate brings a very clear message: "I'm not well." Many volatile compounds that plants exchange with each other carry the same message, and it's amazing that even very different species use identical words to say the same things. Of course, this doesn't mean there is a universal plant language. Rather, it's as if there are different languages with a common root: some meanings remain in all of them, while others are specific to different languages (and therefore to different species).

But back to these volatile messengers, which plants produce and perceive under conditions of stress. Many BVOCs, for example, contain real plant SOSs. Plants produce these compounds after being exposed to stress, whether biological (from fungi, bacteria, insects, or any living thing that significantly disturbs the plant's equilibrium) or nonbiological (for instance, excessive cold or heat, lack of oxygen, the presence of salts or pollutants in the air or in the earth). In all cases, they carry out a surprising function: they warn neighboring plants (or even distant parts of the same plant) of imminent, present danger.

Why? Essentially, for self-defense. Imagine a plant under attack by an herbivorous insect. It immediately releases a molecule to warn nearby plants of the attack. In order to survive the moment of danger, they then mobilize all possible defensive actions, often using remarkable strategies which we will examine later (see "Communication Between Plants and Animals," in chapter 4) but which, to give one example, may include producing molecules that make their leaves indigestible or even poisonous to the insect aggressor. One of the most famous examples is the tomato, which, when attacked by herbivorous insects, releases great quantities of BVOCs in order to alert other plants, even those hundreds of meters away.

But if plants are capable of such effective defensive strategies, why do we need insecticides? And why aren't plants' defenses effective enough to repel all attackers? The answer is very simple. In nature, life is the result of an equilibrium continuously being recreated by the competition between predators and their prey. For every defensive maneuver plants mount against predators, there will always be a new animal strategy to which plants will respond in ever more sophisticated ways over time. This mechanism of continuous improvement is the mainspring for evolution and for all possibility of survival of life on Earth.

Taste

In plants as in animals, the senses of smell and taste are closely connected. In practical terms, the organs responsible for plants' sense of taste are certain receptors for the chemical substances they use as food, substances for which they probe the soil through the exploratory movements of their roots. The "palate" of the plant world proves in this search to be as refined as that of the best gourmets! Maybe you smile at the comparison, but consider that fundamentally, the ability of certain sensitive palates to detect the most minimal ingredients in a dish isn't so different from the ability of roots to identify infinitesimal quantities of mineral salts hidden in many cubic meters of earth.

There is a difference, however. And, as often proves to be the case, it's in the plants' favor. As they perceive minuscule chemical gradients present in the soil, the roots show that their palate is far superior to any animal's! The roots continually taste the soil in search of "appetizing" nutrients such as nitrates, phosphates, or potassium, which they can locate with great precision even in very limited amounts. How do we know this? The plant tells us, by producing many more roots exactly where the concentration of a mineral salt is highest,

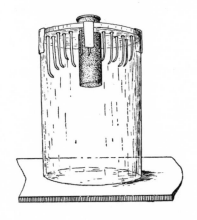

Figure 3-2. Roots grow in the direction of their source of nutriment.

and making them grow until all the mineral salt has been efficiently absorbed.

This behavior is far more sophisticated than it may seem. In fact, the plant, by producing a greater number of roots as a function of the chemical gradient it has identified, acts in advance, investing energy and resources that will yield results only in the future. Somewhat like a mining company investing substantial resources to open new galleries, counting on the future revenues it anticipates earning—another instance of intelligent behavior.

The soil is where we look instinctively for the parts of the plant responsible for taste, because that's where most of the plant world's nutritional resources are found. But many species follow a different diet: the so-called carnivorous plants. Consider the story of *Dionaea muscipula*, the first carnivorous plant discovered by botanists.

On January 24, 1760, Arthur Dobbs, a wealthy landowner in North Carolina and the colony's governor from 1754 to 1765, wrote a letter to the English botanist Peter Collinson (1694–1768), a member of the Royal Society, describing an amazing new plant which had the ability to catch flies:

> But the great wonder of the vegetable kingdom is a very curious unknown species of sensitive; it is a dwarf plant; the leaves are like a narrow segment of a sphere, consisting of two parts, like the cap of a spring purse, the concave part outwards, each of which falls back with indented edges (like an iron spring fox trap); upon any thing touching the leaves, or falling

between them, they instantly close like a spring trap, and confine any insect or any thing that falls between them; it bears a white flower: to this surprising plant I have given the name of Sensitiva Acchiappamosche (Fly Trap Sensitive).

Collinson sent the first samples of this marvelous plant to arrive in Europe to British naturalist John Ellis, who called the species *Dionaea muscipula.* In 1769, Ellis, who had perceived the plant's carnivorous nature, wrote to Linnaeus:

> . . . the plant, of which I now inclose you an exact figure, with a specimen of its leaves and blossoms, shows, that nature may have some view towards its *nourishment*, in forming the upper joint of its leaf like a *machine* to catch food: upon the middle of this lies the bait for the unhappy insect that becomes its prey. Many minute red glands, that cover its inner surface, and which perhaps discharge sweet liquor, tempt the poor animal to taste them: and the instant these tender parts are irritated by its feet, the two lobes rise up, grasp it fast, lock the rows of spines together, and squeeze it to death. And further, lest the strong efforts for life, in the creature thus taken, should serve to disengage it; three small erect spines are fixed near the middle of each lobe, among the glands, that effectually put an end to all its struggles.[2]

---

2. Eighteenth-century spellings have been modernized.—Trans.

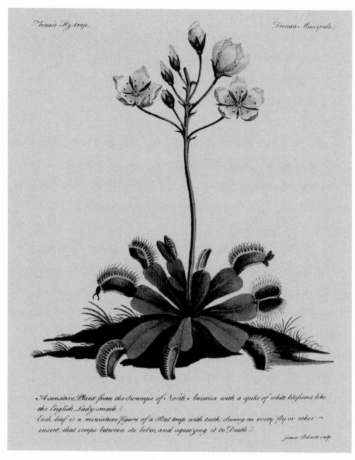

Figure 3-3. *Dionaea muscipula*, a plant native to northern swamps and to South Carolina, in a drawing sent to Linnaeus by the British naturalist John Ellis with his letter of September 23, 1769. This letter contains the first botanical description of a carnivorous plant.

There was no doubt about it: this plant hunted! But Linnaeus didn't think along those lines at all; he rejected Ellis's conclusions and, agreeing with Dobbs's initial assessment, classified the dionaea with the "sensitive plants," which respond to tactile stimuli with involuntary movements.

Today, it seems obvious to us that the dionaea catches insects, but Linnaeus considered it to be the same type of plant as *Mimosa pudica*, which retracts its leaves when touched. The two botanists' conclusions couldn't have been more different: for Ellis, the dionaea was a hunter that preyed on animals; for Linnaeus, it was simply a plant that exhibited an automatic response when touched.

How could these scientists' observations have led them to such disparate conclusions? Ellis, the lesser known of the two, was unaffected by current thinking and limited himself to describing what he had observed and drawing the logical conclusions. But Linnaeus was at the height of his fame, and couldn't detach himself from the idea of the "order of nature" governing relationships among living things, which the entire scientific community of his time accepted. It had such an influence on him that he denied evidence and tried to bend his observations to the theory, at the price of distorting reality. And so, after long research and in the face of irrefutable proof that the plant could catch and kill an insect, Linnaeus refused to affirm (and so to legitimize in the eyes of science) its carnivorous nature, because such behavior by a plant was simply unthinkable.

But what the dionaea could do was visible to anyone: it really did seem able to capture and kill certain insects.

How could this behavior be discounted? Many botanists of that time explained it away by resorting to fancy. They argued that the leaves moved by reflex action (that is, they closed without intending to kill) and that the insects could have freed themselves if they wanted to. If they didn't free themselves, it was because they were old or had chosen to die. Such reasoning seems comical to us, but the scientific community of that time embraced it without hesitation. Any explanation would do if it rejected the possibility that there could be a plant capable of preying on an animal. The hypothesis had to be consigned to the pages of adventure books, where in those days you could almost always find a good man-eating tree.

But how to explain the fact that dionaea never freed the insects before killing and digesting them? And how to interpret the leaf's reopening soon after closing over something tasteless or indigestible?

Reasonable answers to these questions had to wait until 1875, when Charles Darwin published his book *Insectivorous Plants*. Only then did the scientific community begin to speak of "plants that eat insects," a definition that came closer to the truth but was still inexact. For, by Darwin's time, a considerable number of plants had already been discovered and observed that could trap and digest small animals such as mice or lizards. These were hardly insectivores! Dozens of species had been classified as such not because they all preyed on insects but because, in the mid-1800s, *carnivorous* was still considered too strong a word to be associated with plants. Despite the known behavior of many species, especially certain nepenthes, that could catch and kill even small mammals, the existence of plants that consumed flesh was still inconceivable at the end of the nineteenth century.

But why do some plants have this kind of diet? Once more, there's an evolutionary reason. In the humid marshes where these species evolved millions of years ago, nitrogen—essential for life, for the production of protein—was scarce or unavailable. Plants in nitrogen-poor habitats had to find a way to get this vital element that didn't involve their roots and the earth.

How did they do it? They used their aerial parts. Over time, they modified the shape of their leaves, transforming them into traps for insects—those little mobile nitrogen tanks. After imprisoning and killing their prey, these plants proceed to digest it to assimilate its nutrients. In fact, this is a defining characteristic of a carnivorous plant: the ability to metabolize an animal it has caught, by making enzymes that break down the nutrients for the leaf to absorb.

Let's look at the hunting techniques of those ace predators, *Dionaea muscipula* and the nepenthes. Like all great hunters, they start by luring their victims. In the case of the dionaea, the bait is a very fragrant, sugary secretion released on the leaf—which is now a trap—that is irresistible to insects. Contrary to what Linnaeus thought, the plant doesn't have energy to waste and the leaf doesn't snap closed when touched by a hypothetical prey; if it did, it might trap something inedible, or even allow an insect poised on the edge to escape. Instead, the dionaea waits until the animal is right in the middle of the leaf before closing over it, thus avoiding any chance of failure.

On the surface of each of the two parts of the dionaea that make up the death trap, there are three little hairs: they are the trigger that closes the lock. To activate the trap, an insect must do more than touch one hair one time; it must touch at least two hairs, at an interval not longer than twenty seconds. Only then does the plant know it has something interesting and close its leaves. The writhing, trapped animal keeps touching the hairs, which only makes the dionaea tighten its grip. When the animal is dead—and therefore has stopped moving—the leaf starts releasing digestive enzymes, by means of which it will digest the animal almost completely. When the trap reopens, it will still bear the traces of this epic struggle between a plant and an animal; it's not unusual to find on the leaves of a dionaea the exoskeleton (external skeleton) of insects it has caught and eaten.

Those other fearsome predators, the nepenthes, use a different tactic. During their evolution they developed special saclike organs whose edges are sprinkled with sweet, fragrant substances. When the animal, attracted by the odor the plant produces, comes up to the sac to suck the nectar and follow the scent, it slips in and can't get back out. The interior of this trap sac is one of the smoothest things in nature (so smooth that its characteristics are being studied so that they can be imitated technologically). Inside the trap, the unfortunate animal finds digestive fluid, in which it ends up drowning, exhausted by repeated attempts to climb out and save itself. At that point the plant starts digesting its prey, transforming it into a nutrient broth that will slowly be absorbed.

Nepenthes eat not only insects, but also lizards and small

reptiles, and even fairly large mice. The skeletons of their prey are deposited at the bottom of their trap sacs: old hunting trophies, and obscure warnings to the next hapless animals that will be their victims.

Besides being an interesting example of how plants employ the sense of taste, carnivorous plants offer stimulating food for thought about plants' diet. In the first place, contrary to what we've been led to believe, these plants are not few in number. At least 600 species are known, and they each use different kinds of traps and devices to catch various kinds of animals. Plant carnivory, properly speaking, is thus more extensive than was previously thought and involves hundreds of different species. The number increases if we consider plant species that benefit indirectly in some way from the capture of insects. Until a few years ago it was thought that only certain species of plants—those defined precisely as carnivores— had the capacity to digest small animals, thus obtaining from them the nutrients they need. But recent studies have shown that the plant world's use of animal nourishment is quite widespread.

If you've ever looked at the leaves of a potato plant, or a tobacco plant, or even more exotic plants such as *Paulonia tomentosa* (a tree that originated in China and is becoming very common in Europe and the United States), you may have noticed that they often have little insect corpses on them. Why do the leaves of these plants secrete sticky or poisonous substances that kill the insects, if they can't digest them?

The reason is simple and, if you think about it, makes perfect sense: even if the bodies of these insects are not digested right away, they fall to the ground and decompose, releasing nitrogen that the plant needs in its diet; those remaining on the leaf provide nutriment to the bacteria on the plant, which easily absorb their nitrogen-rich waste products.

So quite a few species, without being actual carnivores, also make use of animals to enrich and vary their own diet. The technical term for them is "protocarnivores."

And there are other surprises about plants' diet. In early 2012, a new study described a plant capable of preying on worms with special . . . underground traps! A violet that grows on the very dry, poor soil of the Brazilian Cerrado, it has developed underground leaves that can trap and digest nematodes, a common kind of small worm. The leaves are sticky and the worms that come near get stuck to them; they are then digested, providing a useful supplement to a diet otherwise poor in nitrogen. This discovery is very important because it described an underground hunting technique for the first time, one which might also be found in other species that are typical of very poor soils.

As we've said, to date the number of plants considered to be carnivorous is about 600. But if we added the so-called protocarnivores and other possible underground hunters, we could speak of much greater numbers. And we would get a completely new idea about the diet of plants.

Touch

Two simple questions can help us understand whether plant organisms have the sense of touch: Is a plant aware of being touched by external objects? And can a plant touch with awareness something outside itself, and get information from it?

In the plant world, the sense of touch is closely related to the sense of hearing and makes use of small sensory organs called mechanosensitive channels, found in small numbers everywhere on the plant but with greatest frequency on the epidermal cells, the cells that are in direct contact with the external environment. These special receptors (the mechanosensitive channels) are activated when the plant touches something, or when vibrations reach it. But if the lack of a specialized sensory organ doesn't mean that the plant lacks the corresponding sensory perception, neither does the presence of receptors mean that it does have the corresponding sensory perception, though it may be a good indication.

Does the plant notice that it's been touched? To answer, let's look at the behavior of the *Mimosa pudica*, a particular type of mimosa—called "sensitive," a member of the same group into which Linnaeus classified Dionaea—which retracts its leaves as soon as it's brushed against, as if it were shy (hence the name).

This movement is activated in only a few seconds, and it isn't a conditioned reflex (for example, the leaf doesn't close if it's soaked with water or blown by the wind; it has to be actually touched). So this is true behavior on the part of the plant, but its purpose is perplexing. That it's a defensive strategy seems obvious, but what the mimosa wants to defend itself from isn't at all clear. Some think this sudden closing frightens any herbivorous insects on the leaf; others think the mimosa has evolved this ability so as to look less appetizing to its predators. Which theory is right doesn't

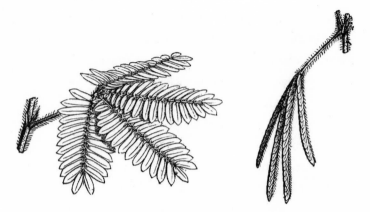

Figure 3-4. *Mimosa pudica* with leaves opened *(left)* and closed *(right)*.
They close instantly in response to precise tactile stimuli.

matter much. What's important is the fact that this plant not only has an extremely developed sense of touch, but can distinguish among different stimuli, and even change its behavior, no longer remaining closed once it learns that a stimulus isn't dangerous.

The first person to perceive the mimosa's extraordinary learning capacity was a giant of science: Jean-Baptiste Lamarck (1744–1829), whose inventions include the word *biology*. He recounted entrusting the following task to his young collaborator Augustin Pyramus de Candolle (1778–1841): to carry some small *Mimosa pudica* plants on a cart through the streets of Paris and describe their behavior.

De Candolle, unfazed by the great Lamarck's request, loaded as many small pots of mimosa as he could onto a cart, and then walked with it around Paris. At a certain point during his walk, de Candolle noticed something unexpected about his unusual cargo. At first, all the mimosa plants had closed their leaves in response to the wagon's bumping along over the Parisian pavement, but soon all the plants reopened; they seemed to have become used to the vibrations.

The explanation for this phenomenon was simple and soon became evident, to de Candolle's astonishment: in a little time the plants had learned that the cart's shaking wasn't dangerous, so they stopped wasting energy by pointlessly closing their leaves.

Observing *Mimosa pudica* obviously isn't the only way to become convinced that plants have a sense of touch. Another powerful example of plants' ability to perceive what's happening on the surface of their leaves or flowers is presented by the carnivorous species. As we've seen, these function like exceptional traps. But when do the traps spring? At the only possible moment, when the insect is on the leaf. Carnivorous plants show that they can clearly perceive when something comes in contact with them, as well as distinguish the kind of tactile sensation that a specific contact provokes.

We know that many plants besides the fierce carnivores show these same capacities. Many flowers also adopt the strategy of closing up when visited by pollinating insects, imprisoning them and freeing them only when they're covered with pollen: a behavior that, again, requires the sense of touch. So now let's ask: if it seems certain that plants have a passive tactile capacity by which they know when something alights on them, do they also have the corresponding active capacity—that is, can they voluntarily touch external objects to get information from them?

To answer this question, the best way to start is to consider the behavior of a root. Each plant, as we have said, has millions of roots (sometimes hundreds of millions), able to penetrate the ground, look for water and nutrients, and then move near them (or away from possible dangerous substances). What happens if, as it approaches a nutrient or water, a root meets with an obstacle, such as a stone? Is its growth interrupted? Does it change course in a preset direction (for example, always going downward or toward the light)? Definitely not.

Laboratory experiments show that the root "touches" the obstacle and continues growing, twisting around it to try to find a way past it. This important function is the task of its extreme point, the root tip, which is endowed with many other extraordinary abilities which we'll take up later, in chapter 5. Thus, the tip touches the obstacle to find out what sort of thing it is, and when it knows, it moves accordingly. For that matter, the roots' ability to do this is fairly intuitive; if they couldn't feel and go around obstacles, how could plants take root in rocky soil?

But roots aside, how does touch function in the rest of the plant?

For the aerial part of the plant, the best example is surely

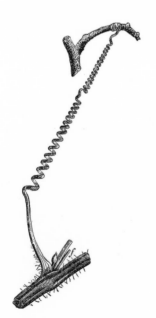

Figure 3-5. *Bryonia dioica* tendril.

the climbing plants (and all tendril-producing plants). Take the pea vine, for example. This delicate little plant produces many sensitive tendrils, which curl up in a few seconds when they touch something, attempting to twine around the object with which they've come in contact. This behavior is found in a great number of plants, which touch the objects around them to find the best one to support them as they grow, and then latch on to it. Could there be any better demonstration of the fact that plants are equipped with touch?

This sense is much in vogue in the plant world; in the last thirty to forty years, ever since statistics of this sort have been kept, it's been calculated that the number of climbing plants is continually growing, surpassing the number of plants that have a trunk.

Imagine for a moment being a newborn plant in the heart of the equatorial forest, where the majority of climbing plants are found. You're little and you face a daunting task: reaching the light. A quick reckoning convinces you that you'll need years and a colossal expenditure of energy to grow a trunk tall enough to get to the light. Scared? There's another option: the short way,

Figure 3-6. A climbing plant: Ipomea purpurea.

the way used by climbing plants. Bona fide slackers, unable to contemplate the sacrifices just described, they take a short-cut to the top, holding on to an already sturdy trunk, and soon reach the light, without any waste of precious energy. This strategy of climbing plants isn't so different from certain human behaviors, wouldn't you say?

Hearing

We come now to one of plants' most controversial senses, one that ignites the collective imagination. Can a plant hear us? And if so, should we talk to it? If you have ever challenged your green thumb, you've asked yourself these questions, and if you've experimented at home, you probably have an answer. Many will say: A plant grows better when I talk to it. Others maintain that talking makes no difference to a plant. It turns out that both answers may be correct, but in order to under-stand why, we need take a step back.

First let's briefly describe the mechanism that enables us to hear, which in a way is what defines the sense of hearing as we know it. The organ dedicated to this sense—in humans as in many animals—is the ear. We know that sounds are actually vibrations, which travel through the air as sound waves and are captured by the earlobes. The earlobes, in turn, conduct the sound waves to the tympanum—the eardrum—which then vibrates, allowing us to translate the waves into sounds. The physical movement of the eardrum's membrane becomes an electrical signal that transmits the information along the auditory nerve to the brain. Thus hearing utilizes air as a pri-mary vector; without air (in a vacuum), the transmission of sound waves would be impossible and we wouldn't be able to hear anything.

We all know plants don't have ears. But this statement of fact shouldn't throw us: by now we've seen that they can see without eyes, taste without taste buds, smell without a nose, and even digest without a stomach. Why should the absence of ears prevent them from hearing?

Here again, evolution played a fundamental role in differentiating plants from us. It gave humans ears at the sides of the head (to capture the sound waves coming from both sides) because, like many other animals, we use air as a vector for sound waves. But plants use a different vector for transmitting sounds: the earth.

How do plants hear? The same way as all the animals that don't have external ears—and there are many. Snakes, worms, and many other animals lack ears, yet they hear. How is that possible?

Their ability to hear comes from the fact that, like plants, they evolved inside an excellent conductor of vibrations. Remember those movies where the Native Americans put their ears to the ground to listen for horses coming from far away? Plants (and snakes, moles, worms, and so on) use the same technique.

The earth conducts sounds so well that ears aren't needed in order to hear; the vibrations can be captured by all the cells of the plant, thanks to the presence of mechanosensitive channels, which we discussed earlier in this chapter with regard to touch. In plants, the sense of hearing also is diffuse, and not concentrated in a single organ as it is in humans. The whole plant is capable of hearing, somewhat as if—below and above ground—it were covered with millions of tiny ears. Thus, like plants' other senses, their hearing evolved in response to the exigencies of living in their environment, half (their most sensitive half) submerged in the soil.

So, like many animals that live in or in close contact with the ground, plant organisms didn't need to develop ears or other specific sense organs, because they hear very well without them.

The functioning of the mechanosensitive channels can be explained with a simple example. Have you ever been in a disco? If you have, you've felt a kind of echo inside your body, somewhere in the belly, produced by the intense vibrations. Even people who can't hear can perceive this type of sound (generally from the bass at full blast), because our bodies vibrate with the sound waves. Well, imagine that for plants the earth is a kind of twenty-four-hour disco. They use the same type of sound reception we've just described, but in a much more sophisticated way.

Over the years many experiments in laboratories and in the field have sought to corroborate the auditory capacity of plants, always with interesting results. Laboratory researchers recently demonstrated how exposure to sounds brings out genetic expression in plants. In the field, a wine-grower from Montalcino, in collaboration with LINV (the International Laboratory of Plant Neurobiology) and with financing from Bose (a corporate leader in the field of sound technology), grew his grapevines to the sound of music for more than five years. The results were astonishing: the vines given musical treatment didn't just grow better than those which had nothing to listen to, they also ripened earlier and produced grapes richer in flavor, color, and polyphenols.

What's more, the music kept insects away, by disorienting them. The use of music made possible a drastic reduction in the use of insecticide, and a new, revolutionary branch of agricultural biology came into being: agricultural phonobiology. In 2011, this project was included by EUBRA—the Euro-Brazilian Sustainable Development Council, promoted by the United Nations—among the one hundred projects that will change the world of the "green economy" in the next twenty years.

Is it really any wonder? For years now, music has been used successfully in the treatment of patients suffering from stroke, coma, epilepsy, and sleep disorders. Music helps us relax or study; it excites and moves us, arouses pleasure or irritation. Even cows seem to enjoy it (classical music), to the point that it has become a requirement for raising the famous Japanese Kobe beef cattle. As for modern music, anyone who practices an individual sport knows that special playlists work a lot better than doping, which is why the wearing of earphones is banned in international competitions, including the New York City Marathon. But though the result is conclusive and has been demonstrated in scientific experiments with plants, as well, we still don't fully understand how music produces this kind of effect. Obviously, plants can't tell one kind of music from another, let alone prefer one of them.

Let's be clear that it's not the kind of music that influences plants' growth, but rather the music's sound frequencies. Certain frequencies, especially bass (between 100 and 500 Hz), promote seed germination, plant growth, and root lengthening, while other, higher frequencies have an inhibitory effect.

More recent experiments on the hypogeal (below the soil surface) rather than the aerial part of the plant have shown that the roots perceive a much broader range of sound vibrations, and that the vibrations perceived can influence the direction of root growth, according to a movement called "phonotropism" (from the Greek *phonos*, "sound," and *trepein*, "turn"). Thus the roots, too, hear and are capable of distinguishing sound frequencies. As a function of the type of vibrations they perceive, they decide whether to move toward or away from the sound source. What use it is to the plant that the roots perceive vibrations? We don't know yet, but the first conjectures on this subject are suggestive and worth mentioning here.

Until a few years ago, it was thought that plants could get information by hearing vibrations transmitted from the soil but, unable to produce sound, they couldn't use that information to communicate with different parts of themselves. But a 2012 study in Italy demonstrated that the roots produce sound, though the way this happens isn't yet clear.

The sounds made by the roots have been provisionally dubbed "clicking," because they characteristically sound like "clicks." In all probability, these tiny clicks result from the breaking of the cell walls—made of cellulose and thus rather rigid—during the cell's growth. Though not intentionally produced by plants, these sounds may be crucially important. Indeed, the discovery opens up new scenarios in plant communication: the fact that roots emit and can perceive sounds would seem to imply the existence of a previously unknown underground communication pathway.

Research results published in 2012, moreover, show that roots manifest organized behavior typical of swarms, involving a kind of communication between the root system of individual plants to explore the earth efficiently, in order to direct growth. A great benefit for those unable to change locations, with limited space at their disposal! We'll talk more about swarm behavior in root systems in chapter 5.

If new discoveries support the theory that the roots can use sounds to communicate among themselves, our idea of plants will again be completely transformed.

. . . And Fifteen Other Senses!

So plants have five senses similar to ours: sight, smell, taste, touch, hearing. From a sensory perspective it would seem that, far from being less endowed than we are, they resemble us. Not so: they're much more sensitive and have at least fifteen other senses that we don't have!

Several of them developed for reasons that are easy to guess. For example, a plant is capable of precisely measuring a soil's humidity and identifying sources of water even at a great distance. It uses a kind of hygrometer (humidity gauge) (from the Greek *hygros*, "moisture," and *metron*, "measure"), very useful for discovering how much water there is in the soil, and where it is. It's easy to imagine why the plant world is equipped with this special capacity which for us, creatures in motion, is much less important. Plants have other extraordinary abilities: for example, they can sense gravity and electromagnetic fields (which influence their growth), and can recognize and measure numerous chemical gradients in the air or in the ground.

Some of these senses are located in the roots, others in the leaves, and still others are diffused throughout the entire plant organism, but what's amazing is the level of sophistication of these green analytical laboratories. Indeed, a plant can locate and recognize trace amounts of chemical elements that are important or harmful to its growth, even when they are several meters away from its roots. Our nose is a much less developed sensory organ! A plant's roots, having perceived a nutrient, turn in its direction and grow until they reach it and can absorb it. Conversely, in the case of pollutants or chemical compounds that endanger both the plant world and the animal world (such as lead, cadmium, or chromium, all unfortunately found in increasing amounts in the soil), the roots move to distance themselves as soon as possible.

Such abilities have been observed for nearly a century and have been studied closely, though never from the proper perspective (the perspective of plants' senses), simply because in our culture, even today, plants aren't considered to be sentient beings (that is, beings capable of perceiving) but passive, insensitive organisms, utterly lacking in every attribute that is typical of animals. And yet, through these abilities, the plant world, indifferent to our meager estimation of it, offers us many kinds of invaluable aid.

We've seen that plants synthesize tens of thousands of molecules, many of which are used in our pharmacopoeia; they

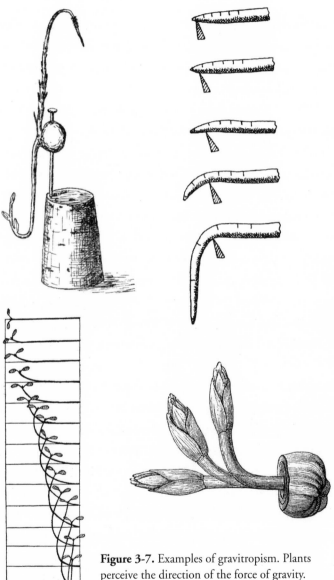

Figure 3-7. Examples of gravitropism. Plants perceive the direction of the force of gravity. The roots direct their growth toward the vector of gravity; the branches and stem develop in the opposite direction.

produce oxygen and provide us with one of our most common construction materials (wood); and in the past, they even produced the energy reserves (fossil fuels) that have sustained our technological growth for centuries. These are indispensable contributions, even before considering that plants are really our only available resource for depolluting the planet.

Take, for example, a substance like trichloroethylene (TCE), an organic solvent used in the plastics industry that pollutes a high percentage of potential water resources in industrialized countries, rendering the water unfit for human consumption. TCE is practically indestructible and can remain undecomposed for tens of thousands of years, a veritable monster of poison and danger; but it is safely absorbed by plants and transformed into chlorine gas, carbon dioxide, and water. In short, it's broken down.

Plants' extraordinary ability to make harmless some of the pollutants most dangerous to humans (and generally produced by humans) and to depollute soils and waters is put to use in various techniques of reclamation being developed in the field of so-called phytoremediation. Although this group of biotechnologies seems to have enormous economic and technological potential to create soil reclamation solutions, their utilization is still in the early stages.

But on that score, at the rate at which we are allowing plant species to become extinct, we are probably forfeiting who knows how many undiscovered solutions and future possibilities for effectively, at modest cost and without environmental impact, depolluting our planet.

CHAPTER 4

Communication in Plants

Imagine a planet where plants have learned to communicate. In this imaginary world they can exchange information and even make themselves understood by animals, including the most complex animals, humans. On this planet, plants have learned to "speak" with animals in their language and can argue persuasively to get the help they need.

They use an information network of other plants and certain animals to extend the reach of their explorations beyond their own organism. They know how to obtain small services and, when necessary, intervention from other species, especially when, unable to change their location, they must defend themselves from herbivorous predators. They also get help with reproducing and propagating themselves in the environment.

Can you imagine such a world, where the most silent, passive, and defenseless organisms we know—plants—influence, and in some ways orchestrate, the lives of animals, from the smallest root worm to human beings? This world already exists: welcome to Earth.

Communication Inside the Plant

Anyone There?

Does a plant communicate within itself? First let's ask a different question: How would having this capacity be useful to a plant? Trying to answer this will help us understand the roots' ability to communicate with the leaves, and vice versa.

With their senses, plants gather information about their environment and orient themselves in the world. Plants are able to measure dozens of different parameters and process a great many data. But for a living organism, unlike a computer, putting information to practical use is more important than collecting infinite amounts of it.

For instance, what if a plant's roots detect that there's no more water in the soil, or what if a leaf is under attack by an herbivore? In such situations, informing the rest of the plant would seem to be essential. Indeed, any delay in transmitting the information could compromise the entire organism's survival. Passing along this news is truly indispensable, but can we really call it communication?

To answer that, let's start by defining what we mean by "communication." Everyone knows the meaning of this word, but sometimes it's useful to redefine a word, even one in common use, to be sure we're all using it the same way. In one of its most common definitions, *communication* refers to the transmission of a message from a sender to a receiver. Communication thus requires three things: a message, a sender, and a receiver. In this elementary communicative model, there's no mention that the two subjects (sender and receiver) must be located in different organisms, and in fact the functioning of our own body—like that of every other living being—clearly demonstrates that communication occurs between different parts of the same organism. For example, if we bang our foot and we feel pain, this is the result of communication between our foot and our brain. In the same way, if we touch something soft and we feel pleasure, this is made possible by the transmission of the tactile sensation from the hand to the brain. Obviously, in any animal, the different parts of the body are capable of transmitting messages.

Communicating is vital to every living being: it allows us to avoid danger, to accumulate experience, to know our own body and the environment. Is there any reason why this simple mechanism should be denied to plants? Perhaps because they don't have a brain? In reality, there's no reason why an organism without a brain shouldn't be capable of transmitting messages inside itself, and in fact, as we'll soon see, plants are quite good at it. It's true that there are some technical obstacles that might seem to make this impossible. Plant organisms aren't equipped with biological structures normally devoted to the transmission of electrical signals, signals which in animals transmit information from the periphery to the central system. In other words, plants don't have nerves. And yet, we've just said that communicating a message is as fundamental for a plant as it is for an animal, and can have equal urgency.

The information that comes from the roots, like that which comes from the leaves, is essential for the entire organism and has to be transmitted rapidly for the plant to stay alive.

The Vascular System of Plants

To transport information from one part of its body to another, a plant uses electrical as well as hydraulic and chemical signals. It thus has three independent systems, which at times are complementary, and which function over both short and long ranges, connecting parts of the same plant that are as close as a few millimeters or as far as dozens of meters apart. Let's briefly look at how these systems work.

The first system, based on electrical signals, is one of the most used, and in practical terms is the same as the electrical system used by animals and humans, though "customized" in certain ways for plants. For example, we've already said plants don't have nerves—that is, tissues dedicated to the transmission of electrical signals, which animals use to conduct nerve impulses. This would seem to present quite a problem: how to transmit signals without having tissues designed for that purpose? Plants have found a very functional solution: for short trips, these signals pass from one cell to another by means of simple openings in their cellular walls, called plasmodesmata (from the Greek *plasma*, "structure," and *desma*, "connection"); for longer trips (for example, from the roots to the leaves), they use the main vascular system.

What? Plants don't have a heart, but they have a vascular system? Yes: like animals, plants are equipped with a hydraulic system which serves mainly to move materials from one point to another within the organism, and which works like a true vascular system, much like our own, except for the fact that it lacks a central pump (that is, it lacks a heart, in accordance with the need to avoid unique organs, which we have already discussed). Thus plants have a circulatory apparatus that permits the transport of liquids from the bottom to the top and vice versa: a sort of system of arteries and veins, called *xylem* when the flow goes from bottom to top and *phloem* when the liquids flow from top to bottom. Xylem (from the Greek *xulon*, "wood") is the conductive tissue principally adapted to the transport of water and mineral salts (but also other substances) from the roots to the crown of the plant, while phloem (from the Greek *phloios*, "cortex") is the tissue that conducts in the other direction, transporting sugars produced by photosynthesis from the leaves to the fruit and roots.

The purpose of this circulation is readily apparent when you consider that the water absorbed by the roots is lost by the leaves in great quantities through transpiration, and so must be continually restored; meanwhile the sugars produced through photosynthesis—the plant's main source of energy— must be continuously moved from the site of production (the leaves) to other parts of the organism.

By means of this complex vascular system, electrical messages circulate smoothly and fairly quickly, as in a tube filled with a conductive solution. Signals that would take a great deal of time to arrive at their destination if transmitted by chemicals can travel in a short time between the roots and the leaves, bringing urgent messages such as those concerning the water status of the soil. Is there only a little water,

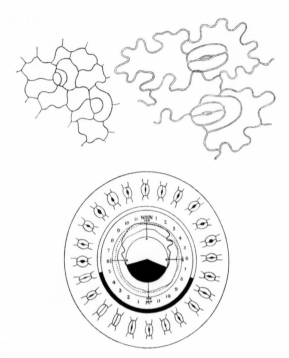

Figure 4-1. The structure of the stomata *(top)*. By means of these small openings on their surfaces, the leaves take in the carbon dioxide they need for photosynthesis and give off water vapor. Under normal conditions, the stomata's cycle of opening and closing *(bottom)* is controlled by the presence and intensity of light.

or a lot? The leaves, with sufficient notice, will adjust to the situation.

The Stomata

Before we come to a concrete example, let's look at the functioning of the stomata (from the Greek *stoma*, "mouth," "opening"), special structures on the surfaces of the leaves (usually on their undersides). These small openings put the inside of the plant in communication with the outside, much like the pores of our skin. Regulating each stoma are two "guard cells," which control its opening and closing based on the current water and light conditions of the plant organism.

The stomata's task is much more complex than it might seem. In fact, balancing the plant's different requirements is far from simple: on the one hand, because carbon dioxide (CO_2, necessary to carry out photosynthesis) enters through the stomata, the plant would seem to have every interest in keeping them open—at least during the daylight hours; but on the other hand, when the stomata are open the plant loses a great deal of water through transpiration.

Every plant has to respond to a real dilemma: keep the stomata open, and through photosynthesis produce the sugars necessary for survival, even if that means losing a great deal of water; or close them, conserving the water it needs but forgoing photosynthesis. It's such a difficult problem that in order to understand how the plant can make the right decision, concepts such as "collective dynamic" or "emergent distributed computing" have been invoked, though they seem a bit out of place applied to plants.

However the plant does it, what's certain is that it reaches a compromise between the exigencies of producing sugar and not losing water, both of which are essential to its survival. Let's look at an example: the summer sun, with its powerful rays, is precious for photosynthesis, as it is for our solar panels. Unlike the latter, however, which produce more energy the more sunlight they're exposed to, a plant must take into account not only light but its reserves of water. This is why during the midday hours—the hottest—it closes its stomata, depriving them of a great opportunity for photosynthesis. By doing so the plant protects itself from the risk of getting too dehydrated.

Imagine a tree (for example an oak or a very tall sequoia), whose roots suddenly notice that there's not enough water available in the soil. Communicating that fact to the leaves is now imperative: if the stomata remained open, continuing to transpire water, the plant could die in a very short time. A very grave danger! So this message is essential for the tree's survival and must travel fast.

To speed it along, as a first option, the plant utilizes an electrical signal, which in a short time reaches the leaves, prompting the stomata to close. At the same time as the electrical signal, chemical/hormonal signals also set out, moving through the vascular system, and taking more time to reach the leaves. These signals move in the same way as chemicals and hormones in our vascular system, but in plants, they are transported by a nutritive solution, rather than in the bloodstream. In a very tall tree, the trip may take several days! But the arrival of the chemical/hormonal signals guarantees that the leaves will get more complete information.

There's a Leak Somewhere!

The hydraulic (vascular) system is also very useful for transporting messages of another sort. Picture a plant organism as a closed system. Have you ever cut into or broken off a branch, a leaf, or the stem of a flower, and noticed a liquid coming out of the wound? The sudden loss of tissue causes a small hydraulic failure in the plant, which communicates a simple but fundamental message to the organism: Attention! There's a leak somewhere! Thus alerted, the plant immediately proceeds to locate the loss and form a scar at the site of the wound.

Thus, as we've seen, the three systems of internal signaling are complementary. They can function over long and short ranges and carry multiple types of information, and each contributes to keeping the plant alive and in equilibrium. From this perspective, too, plants are not so different from us.

Yet despite the similarities, a plant's internal communication pathways have a completely different architecture from those of an animal. Whereas animals are equipped with a central brain toward which all the signals are directed, plants—by virtue of their modular and iterative construction—utilize multiple "data processing centers" which permit them a very different sort of signal handling.

We human beings can't direct a message from our foot to our hand or mouth: all signals, with few exceptions, must first be processed by the brain. Plants, however, can communicate not only from their roots to their crown and vice versa, but also from one root or leaf to another. Their intelligence is distributed! Having no central processing center means that in a plant, information needn't always take the same pathway; instead, it can be transmitted quickly and efficiently right where it's needed.

Communication Between Plants

Plant Language

In our discussion of plants' senses, we saw that they can communicate with each other by means of a real language, composed of thousands of chemical molecules which are released into the air or the water and contain various types of information (see chapter 3). Emitting these molecules is plants' preferred means of communication, just as releasing articulated sounds is preferred by human beings. But we also communicate by means of gestures, facial expressions, bearing, and body language: a system of communication which, though differing from species to species, exists among many animals, especially the higher animals.

And plants? They can communicate with each other, too, by touching (generally with their roots, but sometimes also with their aerial parts) or by positioning themselves in particular ways relative to their neighbors. This is what happens with competing plants during "escape from shade," when they assume different positions relative to each other, vying to win the race to capture light (see chapter 3).

Another example of "gestural" communication is "crown shyness," so named by the French botanist Francis Hallé (b. 1938). This phenomenon, in which some trees tend to avoid touching each other's crown even when growing very close to each other, is not seen in all species, however. Trees usually aren't shy at all about intermingling their crowns. But some species of the families Fagaceae, Pinaceae, and Mirtaceae—to mention a few of the most common—are quite reserved and don't appreciate such interweaving. Just go into a pine woods and look up. The trees manage never to let their crowns touch, but leave a bit of empty space between their own leaves and their neighbor's, thus avoiding contact that we might assume would be unwelcome. Though why and how it happens isn't clear, this phenomenon implies a type of signaling by which the crowns communicate their presence mutually and agree to a sort of territorial partitioning (in this case, of air and of light) so as not to disturb each other.

Plants Recognize Their Kin

Plants interact at many levels and in their interactions exhibit different personalities. Are some species more or less competitive than others, more or less aggressive, collaborative, shy? Sure. But that isn't all. Plants' similarities with the animal world, while not numerous at the anatomical level, are plentiful in the behavioral sphere. This shouldn't be surprising: all living beings have the same fundamental aims, and presumably the means of attaining them are similar in some way, too. Yet despite real affinities between animal and plant behavior, one domain would of necessity seem to be excluded: that of the family. Indeed, plants don't have families. There's nothing like the kind of connection that occurs between related individuals of the same animal species. Or is there?

In the plant world, we don't expect to find the concept of kinship or clan; we tend to associate these notions with very evolved species, such as humans and some other higher animals, but certainly not with plants. And yet . . . plants definitely can recognize their relatives and in general are much friendlier to them than they are to strangers. To understand why plants developed this ability, we should ask ourselves what this trait is useful for. It's an appropriate question because in nature no capacity develops without a reason, including the recognition of familial bonds. Being able to identify the individuals with whom one has strong genetic similarities is important for all species and results in outstanding evolutionary, behavioral, and ecological opportunities. For example: organisms with this capability manage their territory better, defending themselves against enemies without wasting energy fighting against kin; they can avoid reproducing with close relations; and above all they can benefit indirectly from the success of individuals who closely resemble them genetically.

To fully understand these advantages, we need to keep in mind that in nature the main purpose of life is to protect one's genetic inheritance, that is, oneself and one's close kin: parents, brothers and sisters, children. Competing with one of those is such a waste of energy! Much better to cooperate and join forces to overcome adversity, passing our genes to the next generations. From this perspective, the capacity to recognize one's kin is a great advantage; but are we sure that plants behave differently toward other plants based on their degree of genetic kinship?

In the animal kingdom, this process of recognition makes use of the senses: sight, hearing, smell, and in some cases even taste. In plants, it occurs through the exchange of chemical signals released by the roots and probably the leaves (though with regard to leaves, research results aren't yet conclusive).

Plants are stationary, as we've said: a point that bears repeating because this is their main difference from animals. Unable to move from their place of birth, plants clearly evolved as territorial organisms, and their capacity to defend their territory must necessarily be greater than that of any animal. Plants are fierce fighters—and it's not hard to see why. An animal that is in an unfavorable position with respect to another can always go and live somewhere else. A plant doesn't have that option and must resign itself to sharing the resources of its environment with the other beings coexisting in the same area, sometimes even a few centimeters away. But this doesn't mean simply accepting their presence; on the contrary, it means a continuous struggle for one's own space, which must be defended against all intruders. A plant protects its territory by investing much of its energy in its underground part. By producing a great many roots, it occupies the soil like a military force and claims possession of it against its neighbors. But not always: if neighboring plants are part of the same clan, and thus relatives, there's no need to compete, and the roots can be kept to a minimum for the benefit of the aerial part.

In 2007, a simple but important study shed light on this type of familial behavior. The experiment consisted of growing thirty seeds from the same plant in one pot, and in a second pot, identical to the first, growing thirty seeds from different plants. Observing the behavior of the young specimens growing in the two pots led to the discovery of several evolutionary mechanisms formerly thought to be present only in animals. The thirty plants from different mothers behaved as expected, developing a greater number of roots in an attempt to dominate the territory and assure themselves sufficient food and water, to the disadvantage of the other plants. But the thirty plants from the same mother, though they too found themselves coexisting in a restricted space, produced many fewer roots, advantaging the plants' aerial growth. In their case, what was observed was noncompetitive activity linked to their genetic proximity. This was a fundamental discovery: it replaced the traditional view that plants would adopt a stereotypical and repetitive mechanism (neighboring plant = necessity for defense and competition for territory) with a much more complex estimation that takes into account different factors, including genetic kinship. The plant, it turns out, checks out a potential rival before attacking or defending, and if it discovers a genetic affinity, instead of competing it chooses to cooperate.

Selfishness or Altruism: Which Is More Useful?

In evolutionary terms, which is more rewarding—behavior we call "selfish," or behavior we call "altruistic"? The jury is still out on that question. Of the innumerable simulations and models that have been constructed, none were ever thought to be applicable to the plant world. The discovery that plants adopt altruistic behaviors toward kin is revelatory because it opens up two possibilities, both revolutionary: either plants are much more evolved organisms than we thought, and thus are altruistic; or altruism and cooperation actually are primitive forms of life, where pure competition had always been thought to rule, with victory going to the stronger. In either case, communication between one plant and another by means of their roots would have a precise evolutionary purpose: distinguishing strangers from relatives. And enemies from friends.

Continuing our discussion of the behavior of roots (whose special capacities we'll examine in detail in the next chapter), they seem able to communicate not only with the other plants, but with all the organisms of the so-called rhizosphere (from the Greek *rhiza*, "root," and *sphaira*, "sphere")—that is, the part of the soil with which they are in contact and which hosts a great many other forms of life. It's a fairly common misconception that the soil is an inert substrate; on the contrary, the soil is a living and densely populated environment. Microorganisms, bacteria, fungi, and insects form a special ecological niche that stays in equilibrium because of communication and collaboration with plants.

One very common instance is that of the mycorrhizae (from the Greek *mykes*, "fungus," and *rhiza*, "root")—special forms of symbiosis occurring in the subsoil between the vegetative part of fungi that we commonly eat or see in the woods, and the roots of many species of plants. In certain cases, the fungus forms a sort of sleeve around the plant, penetrating inside its cells. This kind of symbiotic association is called "mutualistic" because it's useful to both living organisms: the fungus provides the roots with mineral elements, including phosphorus (always hard to find in adequate amounts in the soil), and in exchange receives some of the sugars produced by the plant through photosynthesis, which it uses as a source of energy.

But in this relationship, which seems so convenient, there can be some rude surprises. The problem is that not all fungi have collaborative and peaceful intentions: some are pathogens and will attach to the root to get nourishment, destroying it in the process. So the plant needs to be able to distinguish what type of fungus is attempting to come into contact with it, and act accordingly. But how can it tell a friendly fungus from an enemy? Recognition results from a real chemical dialogue between the fungus and the root, which exchange signals to clarify their respective intentions. If the plant perceives that fungi have hostile designs, it will initiate hostilities. On the other hand, if after proper introductions it recognizes that this is a well-intentioned mycorrhizal fungus, then it will allow this symbiotic relationship which is so useful to them both to be established.

A Bacterium for a Friend

Another example of mutually beneficial symbiosis based on plant communication is the relationship between legumes and nitrogen-fixing bacteria. Along with a few others, these microorganisms have a capacity that's extraordinarily useful to living things: they fix atmospheric nitrogen, transforming it into ammonia (NH_3) by breaking the tight bond between the two atoms making up a molecule of gaseous nitrogen (N_2).

Nitrogen is the principal element of a soil's fertility (which is why many fertilizers are based on nitrogen compounds), and although the air we breathe is 80 percent gaseous nitrogen, this gas is inert and cannot be utilized by plants or any other living beings, except a few microorganisms such as these same nitrogen-fixing bacteria. These bacteria, as we were saying, transform gaseous nitrogen into forms of nitrogen, such as ammonia, that are easily absorbed by plants. In effect, they're natural fertilizers! In return, the bacteria find inside the roots an ideal growing environment and sugars in abundance—another tale of mutual benefit, again based on communication and recognition. Not all bacteria, in fact, are welcomed by plants; many are frightful pathogens against which plants construct unassailable barriers. Nitrogen-fixing bacteria, before being welcomed, initiate a long and complex chemical dialogue with the roots. This "conversation" unfailingly begins with the release of a signal from the bacteria that's like a password, called NOD factor (nodulation, or nod, factor), whose recognition by the plant is the first step in the plant's granting the bacteria free entry to the root.

Examples of symbiosis like the one just described are all based on close communication between the symbionts (as the two partners in symbiosis are called—in our example, the bacteria and the legume) and couldn't occur if collaboration between living organisms were not a long-established practice. In fact, these phenomena aren't confined exclusively to the plant world or lower organisms. On the contrary, some of these symbiotic relationships have become so established and important that they're the basis of our own life.

Let's look at an example: Mitochondria are the energy centers of our cells (or rather, of all animal and plant cells). Without these organelles situated within each cell, the existence of higher forms of life would be inconceivable. Well, recent studies suggest that mitochondria, too, are the result of

symbiosis; in this case, symbiosis between the cells and primitive bacteria endowed with a powerful oxidative metabolism
(in other words, capable of producing energy). The bacteria
and the cells enter into a symbiosis which benefits them both
(the bacteria produce energy for the cell, and in return obtain
everything they need to survive), and at a certain point the
bacteria are incorporated into the cells. There's a great deal
of evidence to support this theory of the symbiotic origin of
mitochondria. First of all, the mitochondrion exhibits many
of the characteristics typical of bacteria, including a membrane very similar to theirs; then—again like bacteria—it has
a closed, circular, double-helical DNA; and finally—and this
is the most important evidence—like bacteria, mitochondria
replicate independently of the organism that contains them.
Several studies have clarified the fundamental importance
that these formerly symbiotic cells have had in the evolution
of complex forms of life.

Symbiotic relationships thus are fundamental for all forms
of life on the planet, and for our own existence: if we could
learn to direct some of them, the results could be spectacular.
For example, if we could transfer the symbiotic association
between plants and nitrogen-fixing bacteria from just the
legumes (which include, among others, soybeans, chick peas,
lentils, peas, and beans) to all food crops, we could change the
face of agriculture forever.

Imagine: no more nitrogen fertilizers, no more pollution
of the soil, groundwater, rivers, and oceans, no more algae
on the Adriatic, but instead increased crop productivity and
the possibility of feeding the world without polluting it—a

dream we would be wise to invest in so that research efforts can be multiplied, and because we need it to come true as soon as possible in order to avert disastrous consequences.

From the end of the Second World War until today, the productivity of crops and soils grew continuously, thanks mostly to the so-called Green Revolution of the 1960s. With the use of chemical fertilizers and the creation of new, more productive and resistant plant varieties, this great agricultural modernization led to the cultivation of new land and increased yields on land already under cultivation, ensuring extraordinary production increases for many years.

But now, the upward trend in crop yields has been interrupted. For the first time in sixty years, land under cultivation not only isn't increasing, it's actually decreasing because of climatic changes, while the world's population continues to grow.

How will we feed ourselves? Finding the means to create a new "Green Revolution"—a system that will enable renewed increases in crop productivity in a way that is environmentally sustainable—is one of the priorities of the decades ahead. That's why the possibility of extending symbiosis with nitrogen-fixing bacteria to all crops could truly be a breakthrough. Plant communication would help us feed the world!

Communication Between Plants and Animals

Mail and Telecommunications
Plants' internal communications, as the business world would call them, function very efficiently. But how do plants manage communication with the outside world?

Because plants can't move from the place where they were born, they need help to receive and send messages outside themselves, as well as small objects such as pollen and seeds. So they've adopted a sort of postal system. Sometimes they use air as a mail carrier, sometimes water, but most often they use animals, especially for very delicate operations such as defense or reproduction. Would any of us send a sensitive message via a bottle or a paper airplane? It would be much better to have an animal take our message and deliver it to the recipient (think of the carrier pigeons that humans used for this purpose for centuries). But how do plants persuade insects and other animals to be their pony express?

In the section on "Honest and Dishonest Plants" later in this chapter, we'll discuss at length the various modalities of mating among plants, and their ways of persuading animals to help them with pollination and propagation. First, let's look at the other circumstances in which plants decide to seek animals' help, starting with the most common: defense.

Help! Send Reinforcements! (Plant Defense Systems Based on Communication)

Let's say an insect settles on a plant's leaf and starts to eat it. The plant, noticing that it's being attacked, immediately mounts a defense strategy. First it identifies the insect aggressor; only then—by knowing what is attacking it—will it be able to defend itself adequately.

Most often, the plant uses chemical weapons, producing special substances that make the leaf unappetizing, indigestible, or even poisonous to the herbivore. To avoid a waste of precious energy, the production of these "deterrent substances" occurs exclusively inside the leaf that's under attack, and in the leaves next to it, in anticipation that this initial act of deterrence may be enough to discourage the insect. Why should the plant waste energy mobilizing all its resources if it can find a local solution?

Each choice a plant makes is based on this type of calculation: what is the smallest quantity of resources that will serve to solve the problem? In fact, this calculus and the strategy associated with it often succeed: in our example, the insect will taste one or two leaves and then, repelled by the new flavor, will leave the plant and move to another one. Victory!

By sprouting new leaves, the plant easily repairs the small damage it has experienced, and it won't be much affected by the loss; as we know, its organism is constructed in such a way that the removal of even substantial parts doesn't compromise its functioning or its survival. In our example, the plant's reaction to aggression was mild—we could almost say benign.

But if the insect keeps feeding on the leaves despite their foul taste, or if other culinary adventurers come to the table, the plant will be obliged to use more drastic strategies: in some cases, it starts producing "deterrent" chemical substances on all its leaves and releases volatile chemical signals into the air that alert neighboring plants to do the same. In other cases, the plant may choose to . . . call in reinforcements!

The Enemy of My Enemy Is My Friend
Each new day witnesses the continuation of a 400-million-year-old battle for survival between herbivorous organisms

and plants. The most important group of herbivores unquestionably are insects, which find in plants an enormous variety of habitats and ecological niches as well as, obviously, a great deal of food. This endless conflict exerts an extraordinary selective pressure, which shapes the evolution of both plants and insects and controls their distribution in space and time.

To deal with the attacks and damages inflicted on them by insects, plants have developed a series of articulated defense strategies; insects, for their part, have not stood idly by, but instead have devised ever newer and more effective plans of attack. It's a sort of perpetual arms race, resulting from the coevolution of plants and herbivores, natural enemies who, through their encounters, have come to know each other very well.

Have you ever read, perhaps on a package of salad, the phrase "Produced by Integrated Pest Management"? This means the growers have chosen to reduce their use of pesticides in cultivating these greens, by introducing into their fields natural enemies of the herbivorous insects that normally attack lettuce. Rather than spray their fields with pesticides, the growers have relied on the enemies of the greens-eating insects to hunt them or at least keep them occupied, away from the plants—a very clever technique, though difficult to master because of the necessity to keep the insect populations in equilibrium. The way it works could be summed up in the phrase: "The enemy of my enemy is my friend."

Many plants use just this strategy to defend themselves: they ask the enemies of their enemies for reinforcements, appealing to them by producing volatile chemical substances and afterward repaying them for their help—behavior that brings excellent results without much energy expenditure.

Let's look at an example: the lima bean. When attacked by especially voracious mites (*Tetranychus urticae*), the lima bean releases a mix of volatile chemical substances that attract a different mite, a carnivorous one (*Phytoseiulus persimilis*). This mite specializes in attacking "vegetarian" mites and soon exterminates the entire population—another amazing instance of collaboration between an animal species and a plant species, one that depends on the lima bean's even more amazing ability to recognize its aggressor and then call on one of its aggressor's biological enemies to come to the rescue.

How many animals are capable of such an evolved strategy? A great many plant species are: among them corn, tomato, and tobacco, to name just a few.

The Case of Corn

We've seen how a plant behaves when its leaves are assaulted by an herbivore. But what happens when the attack isn't launched against the leaves, but against the roots? An emblematic example is corn. In the United States, corn crops have been decimated for years (with hundreds of millions of dollars in damage) by *Diabrotica virgifera*, an insect that deposits its larvae on the roots, killing the young plants, which are incapable of defending themselves. So in terms of self-defense, corn seems to be a flop among plants. And yet . . . this isn't its fault!

The oldest European varieties of corn and the wildest—each the fruit of a very long selection process, very different from what we grow today—were perfectly capable of defending themselves from the diabrotica's attacks. It was we who indirectly and unknowingly—by a process of selecting new varieties in order to obtain higher-yielding specimens with larger cobs—selected plants that couldn't defend themselves. The old varieties of corn, when attacked by *Diabrotica virgifer* depositing their larvae close to the roots, produced a substance called caryophyllene whose sole function is to call to the aid of the plant small worms (nematodes) that love to gorge on diabrotica larvae. Eating them, the worms freed the corn of the parasite.

Our unwitting mistake, which led to the selection of defenseless corn varieties, has cost us dearly: it's been estimated that worldwide, losses due to this insect amount to around a billion dollars per year! For decades, the diabrotica has been corn's scourge, and huge sums have been spent combating it, with tons of insecticides spread in the atmosphere. Only through genetic engineering has corn's original capacity been restored: the gene regulating caryophyllene production has been reintroduced in modern varieties, after being borrowed from oregano. In short, in order to restore an innate characteristic in corn, we had to create a genetically modified (transgenic) plant.

Plant Sex

One of the moments when a plant needs to communicate most, especially with animals, is during pollination. This period, which we might call plants' breeding season, is a crucial phase in their life; their chance for successful reproduction depends on it. Obviously, each plant is different; yet for most species, from geraniums to oaks, certain common rules apply. For example, in many plants fertilization requires that their pollen (the plant equivalent of male semen) be transported from one flower to another. But before we consider the mysteries of communication between plants and animals, let's step back to see how plant reproduction works.

First let's distinguish between plants that are autogamous (from the Greek *autos*, "oneself," and *gamos*, "sexual union") and plants that are allogamous (from *alios*, "other," and *gamos*). Autogamous plants use an "autarkic" method, pollinating themselves through the passage of pollen from the stamen (the male reproductive organ) to the pistil (the female reproductive organ) of the same flower. In allogamous plants, in contrast, the pollen must be transported from the anther (the tip of the male organ, containing pollen granules) of one flower to the stigma (the pollen-receiving part of the female organ) of another flower belonging to a different individual of the same species: thus allogamous plants practice "cross pollination."

Another distinction among plants has to do with the location of the sex organs. In this respect, plants are broadly divided into three categories: hermaphroditic, dioecious, and monoecious.

The first and by far the largest category is that of the hermaphrodites, whose flowers have both male and female sex

Figure 4-2. Pollen granule. In plant reproduction these granules are the male gamete (male seed).

organs. In hermaphroditic plants, theoretically, each flower can fertilize itself on its own because it has both reproductive apparatuses. Thus, according to the definition given above, a hermaphroditic plant is autogamous. Self-fertilization is a great convenience and is practiced by a number of plants, especially grasses (such as wheat or rice). These, along with certain species of orchids, violets, and carnivorous plants, are actually cleistogamous (from *kleistos*, "closed," and *gamos*), that is, they pollinate themselves before the flowers even open.

Although in theory self-pollination is possible in all plant species with hermaphroditic flowers, in practice it occurs infrequently, prevented by a series of physical or chemical barriers. Why is this, since we've just pointed out how convenient self-pollination is to the plant?

The reason is easy to see, if we consider that self-pollination is the plant equivalent of consanguineous procreation (procreation between close relatives) in the animal world: evolution discourages this kind of reproduction because it reduces the appearance of new genetic combinations. So plants have evolved a series of special mechanisms for avoiding self-pollination, such as the maturation of the male and female apparatuses at different times within the same individual.

Another category of plants is made up of dioecious species (from *dis*, "twice," and *oikia*, "house"), which have single-sex flowers on individuals of different sexes, so that for every plant there are "male" and "female" specimens. Among these is *Ginko bilobo*, a very ancient tree that can be considered a species of living fossil, as well as laurel, butcher's broom, yew, nettle, and holly.

Finally there is a third category, monoecious (from *mono*, "one," and *oikia*, "house") plants, such as oak trees and chestnut trees, which have separate male and female flowers on the same individual.

No matter which category plants belong to, when they are blossoming, they need reliable vectors to transport pollen from one flower to the pistil of another. Each species achieves this in its own way: some rely on a physical vector—the wind; others on animals. The former (known as anemophiles, from the Greek *anemos*, "wind," and *philos*, "friend") on one hand face no complications resulting from the need to attract animals or to have anything to do with them. On the other hand, they must solve the problems resulting from their choice of a totally unselective vector: once airborne, the pollen may land on another plant, an automobile, the ground, anywhere. Thus, in order for the errand to have a chance of success, they must produce a great number of flowers that release incredible quantities of pollen into the air (causing, among other things, many miserable springtime allergies). In terms of energy, as you can easily imagine, this isn't a very efficient system, and it's used mainly by ancient species such as the gymnosperms (so named because the seeds aren't protected by an ovary, from the Greek *gumnos*, "nude," and *sperma*); but also by many more-recent angiosperms, such as olive trees.

Most modern plants, however, rely on animal vectors, which are much more precise in the processes of gathering and delivering pollen. The animals most commonly used as vectors are insects—prized helpers charged with so-called entomophilous pollination (from *entomon*, "insect," and *philos*,

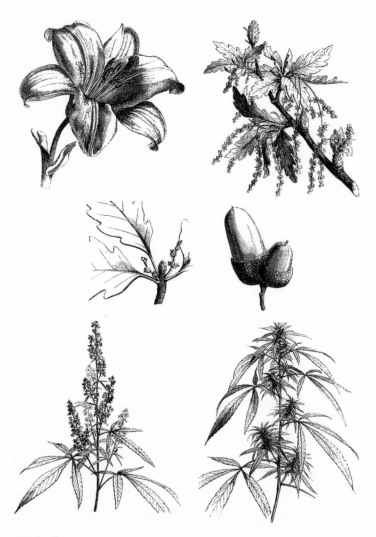

Figure 4-3. Location of sexual organs in plants. In plants with hermaphroditic flowers, such as lilies *(top, left)*, the male and female organs are found on the same flower; in monoecious species, such as the oak *(top right and center)*, they are separated but on the same plant; in dioecious species, such as hemp *(bottom)*, male and female flowers are found on different plants.

"friend"). But insects aren't the only animals to which plants entrust this delicate delivery. In zoophilous pollination, the vectors are various types of animals (*zoa* in Greek); in ornithophilous pollination, the vectors are birds (*ornites* in Greek) such as hummingbirds and parrots; and in chiropterophilous pollination, the vectors are bats (*cheiropteroi* in Greek), which are used to carry pollen from many American desert cacti, such as the Seguaro cactus.

Recently, in a liana native to Cuba, the *Marcgravia evenia*, circular leaves shaped like real satellite dishes have been described, whose only purpose appears to be to signal the presence of flowers to bats' sonar. They're a little weird-looking; but having chosen animals that can't see well to be their pollinators, why shouldn't the plant help them find its flowers?

Figure 4-4. Cactus. These plants are adapted to hot, dry climates. To survive, their flowers open only at night. Many cacti use bats as vectors for pollination.

Other forms of zoophilous pollination make use of reptiles (different *Pandanus*, for example, use a species of gecko), marsupials, and even primates. Plants have recruited every type of animal to the ranks of pollinators!

The World's Biggest Market

Think of plant pollination as a huge market. We have buyers (insects), goods (pollen and nectar), sellers (plants), and even . . . advertising (the color and fragrance of flowers)!

In the plant world as in the animal, no one does anything for nothing, and in the great pollination "market" there's a real exchange of goods and services. Whoever wants a good or service pays for it. Insects pay with their labor. But plants use a unique currency: nectar, a sweet and very energy-rich substance that animals adore. In fact, it now seems certain that nectar is produced by plants for this sole purpose: as a commodity to be purchased with the transport of pollen.

To generalize a bit: an animal of some kind or other (a lizard, a bat, or a bee) comes to the flower to eat or collect nectar, and in doing so gets covered with pollen, which it will bring to another flower. Obviously, not just any flower will do: it must be the same species of flower as the one from which the pollen was brought. Just as we can't cross a cricket with a hippopotamus, we can't cross an apple with a violet. Pollen transported between two different species would be wasted; but how is an animal carrying pollen from a certain species persuaded to visit other flowers of the same species? What can induce such species loyalty? The much easier course would be to get nectar wherever it is, simply by visiting plants that are closer together, regardless of the species they belong to. But this isn't how insects behave; all day long, they stay loyal to the first species they visit in the morning.

To describe this unusual behavior, which is one of the factors on which the whole process of pollination, and so plant reproduction, is based, entomologists use the term *site fidelity.* This phenomenon has been completely underestimated by researchers, who've yet to come up with a convincing hypothesis to explain it. Botanists and entomologists know very well that a bee will stay all day on the same species of flower it visited in the morning. Yet, incredibly, they have no plausible explanation for this behavior; the theories offered are few and inadequate, generally attempting to show that such fidelity is a practical matter for the insect. From all evidence, on the contrary, this behavior isn't practical at all!

But if, instead, we look at the problem from the point of view of the plant, such fidelity is of paramount importance. A plant would have no interest in producing nectar if its pollen were to end up misplaced. This simple consideration would suggest that it's the plant that seeks and wins from insects their "site fidelity." Just how it does this, however, has yet to be discovered.

Honest and Dishonest Plants

Such mysteries aside, pollination seems at first glance to be an honest and transparent business: the pollen carrier is compensated with nectar. Yet something can always go wrong. Every market has its honest and dishonest merchants; some deal fairly with their customers, others cheat them. It's no different with plants. Some seem to be scrupulously loyal, while others get what they want through disguise and subterfuge, even to the point of imprisoning insects that help them. Some stop at nothing to get what they need.

Let's start with lupines. These legumes produce a great many small flowers, and they have a problem: how to prevent bees from revisiting the same flower. If an industrious hymenopteran completes its task the first time it visits a flower—gathering nectar and getting covered with pollen which it then transports to another flower—a second visit to the first flower by the same or another bee would be a complete waste, since there's no more pollen, nor nectar to suck. To avoid this mishap, which in addition could leave some flowers unpollinated, the lupine adopts a straightforward and very effective strategy: it changes the color of the petals of flowers that have been visited (and so have no more pollen or nectar), tinting them blue. In this way, it lets the insects know there's no more available nectar and they should go on to another flower. This strategy is very helpful to the pollinators and at the same time works very well for the plant, which has a more successful pollination.

But as we've already said, not all plants are the same. So, while lupines show exemplary trustworthiness in their dealings with their animal partners, other plants use different strategies to pursue their goals, and are just as successful. The most notorious examples are orchids: according to some estimates about one-third of existing orchid species, to guarantee themselves a successful pollination, use strategies that in human terms could only be described as fraudulent. These plants make use of insects, too, but by deceiving them and getting them to transport pollen without the insect's receiving any benefit in return. We hasten to say that we can't properly speak of honesty and dishonesty in nature, but even so, it's interesting to see how orchids manage to deceive insects. These plants are among the most mimetically talented of all living organisms. Usually, when we speak of mimesis we think of animals such as chameleons or walkingsticks. But their considerable mimetic abilities are as nothing compared to what an orchid like *Orphyrs apifera* can do.

Its flowers are able to mimic perfectly the shape of the female of certain nonsocial hymenoptera (which are similar to wasps and bees but don't live in colonies). And that isn't all: besides the female insect's shape, the orchid imitates the consistency of its tissues, its surface (including the fuzz on its body), and of course also its scent, secreting pheromones identical to the ones produced by females ready to mate. Thus the plant performs a triple mimesis, duplicating the form and colors of the female body (deceiving sight), its fuzzy surface (deceiving touch), and its special scent (deceiving smell). So perfect is the resemblance that the male insects can't help going astray. The captivated males are unfailingly duped by these seductive flowers; the mimesis is so lifelike, they even end up copulating with the flower.

The deception is realistic enough to outstrip reality, so that when the orchids are in bloom, these hymenoptera prefer to copulate with the flowers even when their own females are available! Then, when the male insect starts copulating with what it takes to be a female of its species, a mechanism springs that showers its head with pollen packets, tiny boxes full of pollen that the fornicator won't be able to get off itself for some time to come, and that it will perforce bring with it when it visits (and pollinates) the next flower. Just which one is running this show, the plant or the insect, seems absolutely clear.

Money Doesn't Stink (Or Does It?)

If orchids have become master illusionists by perfecting their cheating techniques, many other species, though not reaching their level of perfection, practice the art of deceit on insect victims, too. Take, for instance, the *Arum palaestinum*, a plant native to Israel, Jordan, Lebanon, and Syria (and found in northwestern California, where it has been introduced). For its pollinator, this plant makes use of the drosophila, the common fruit fly, by means of a curious deception. To falsely attract the fly, the *Arum palestinum* produces an aroma that the fly can't resist: that of fermenting fruit. Attracted by this odor, the drosophila blithely enters the inflorescence, which closes after it and holds it prisoner usually for an entire night. During these hours of imprisonment the fly, vainly trying to escape, keeps flying, walking, and sliding around, and gets completely covered with pollen. When the inflorescence opens, the insect finally escapes, but it usually doesn't go far. Drawn once again by the irresistible aroma of fermenting fruit, our hero soon slips into another arum flower which, after imprisoning it in turn, uses the pollen covering the drosophila to pollinate itself. In this way, through deceit, the arum gets what it wants (to be pollinated); the drosophila does the job of pollen transport, but gets nothing for it.

In reality, there are a great many such examples based on the olfactory attraction of insects. An odd case, which could rightly be called macroscopic, is that of the *Arum titanium* (*Amorphophallus titanum*), often called the corpse flower, the plant which has the largest inflorescence in the world. This botanical superstar, whose blooming every year brings curiosity-seekers to botanical gardens, has chosen an efficient but unappealing pollinator: the carrion fly. To attract it, the plant perfectly duplicates its favorite odor: the stench of rotting flesh!

Plants certainly have great manipulative capacities—at this point could anyone doubt it? But let's try for a moment to put ourselves in their places, and ask a possibly unsettling question: Which animal vector is the most efficient for a plant? Without a doubt, it's humans, who guarantee the reproduction, survival, and propagation of certain plant species, to the detriment of others.

From a plant's perspective, it would be worth the trouble to make friends with these strange bipeds, and even to benefit from their services! Are we sure they haven't used their manipulative skills with us too, creating flowers, fruits, fragrances, flavors, and colors that please our species? Maybe plants produce them for just that reason: because they please us, who in return propagate them throughout the world, care for them, protect them. When we think of the marvelous gifts plants give us—from perfumes to the wonderful, multicolored forms that have inspired so many artists—let's not be too surprised at our good fortune. Nobody does anything for nothing, and at least for certain species, we're the best possible allies on the planet.

A Very Special "Postal System"

With respect to plant reproduction and especially the transport of seeds, there are many more examples of plants' ability to communicate with animals. The formation and subsequent dispersal of seeds is the last fundamental phase of plant reproduction. For every plant, dispersing its seeds successfully in the environment (recall that the seeds contain the embryo of the new specimen) is vital for at least two excellent reasons: the first—to extend over the largest possible territory—is a fundamental principle of life for every species; the second— to spread the seeds far from the mother plant—allows the plant to avoid sharing resources in a restricted area, which could soon become deficient in nutriments and therefore unable to assure the sustenance of progeny. Think, for example, of plants that rely on the wind for seed dispersal, like the famous dandelion, which we delight in blowing on and robbing of its seeds. The flower is an extraordinary feat of engineering, constructed so that its minuscule seeds fly away on the slightest puff of wind, sometimes traveling for miles. The seeds of another anemophilous plant, the linden tree, can fly for a long time on their single wing, floating on a light breeze. But what interests us now is plants' use of animals to spread their seeds. A great many species of animals form "business" relationships with the plant world, from birds to fish, from mice to ants, to numerous mammals, even very large ones.

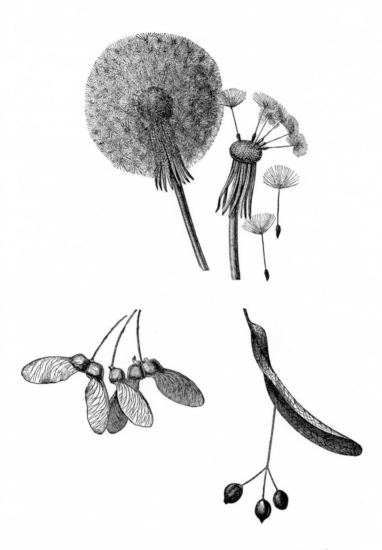

Figure 4-5. Examples of anemophilous plants, characterized by "flying seeds." To disperse their seeds in the most efficient way possible, plants that use the wind as a vector have evolved seeds equipped with special flight systems. Shown here: the parachute system of flight—seen in the dandelion *(top)*; the winged system—seen in maples *(bottom left)*; and the single propeller, seen in the linden *(bottom right)*.

To explain how this communication works, we have to begin with fruits. Fruits are actually plants' means of inducing animals to transport their seeds, in the same way that nectar is used in pollination. Whether it's apples, coconuts, cherries, or apricots, the delicious, sweet pulp of fruits serves two main purposes: to protect the seed until its maturation is complete, and to reward the pony express that transports the seed.

Fruits: "Gift Packages" for the Mail Carriers

All fruits, not only those we consider edible, are produced to hold the seed and also, usually, to entice animals. Eating a fruit in most cases actually means eating the seed too, and transporting the seed far from the plant that made it—by expelling it elsewhere. This is one of the most efficient ways of guaranteeing seed dispersal.

In countries with a temperate or tropical climate, one of the most common vectors of seeds is birds. To see how this kind of communication between plants and animals works, let's use the cherry tree as an example. During pollination, this tree produces flowers of a beautiful white color that seems purposely designed—as in fact it is!—to attract bees, which can see it very well and thus find the flowers more easily. But bees don't see red. The fruits (cherries) have that color not to attract bees, but to attract birds. Indeed, red stands out very well among leaves even from a great distance, and thus can easily be seen by a bird in flight.

Attracted by this winking color, the bird finds a cherry and eats it, seed and all. Then it takes off again and at a certain point somewhere, with its feces (an excellent fertilizer), it expels the seed—a very efficient transport system, convenient both for the plant, whose seed is thus spread far from the mother plant, and for the animal that is fed. But take note! The cherry turns red only when the seed is mature; until then it's almost invisible to birds because of its green color, which keeps it concealed among the leaves.

Every plant tends to protect its own fruits until they ripen. Unripened fruit, in fact, is full of toxic chemical substances that are astringent or even unpleasant, substances the plant uses to defend itself from animal predation before the seeds are mature. To achieve this, the plant sometimes has to use molecules with a very high level of toxicity. Such is the case with ackee (*Blighia sapida*), a wild plant native to Africa that also grows in the Caribbean. This plant produces fruits, delicious when fully ripe, that are eaten by many Central American populations. But you have to make sure they're ripe; when still green they contain high levels of hypoglycin, a substance which, if ingested, causes severe poisoning, with symptoms typical of hypoglycemia: coma, convulsions, delirium, toxic hepatitis, acute dehydration, and shock. Ingestion of unripe fruits causes about twenty human deaths every year.

Obviously, birds aren't the only animals used as seed vectors by plants. Another important group is represented by frugiferous (fruit-eating) monkeys, which are a significant resource for seed dispersal. Finally, there are more unusual animal vectors. In Amazonia, the *Colossoma macropomum*, a large freshwater fish, performs this role in an extraordinary way. During the rainy season, when river flooding results in the formation of almost 100,000 square miles of temporary lakes, the colossoma eats the fruits of many plants and excretes their seeds as far as hundreds of miles away—an interesting dispersal strategy that was only recently discovered.

And then there are the ants. Among the foods these insects eat are small fruits, which they don't consume on the spot but bring to the anthill, putting them "in the pantry" to be eaten later. This is an especially gratifying custom to the plants, which thus have two requirements met at one time: not only are the seeds transported far from the mother plant, but they go almost directly underground, to an ideal place for their future germination. The help of ants, in short, is truly precious, and it's not surprising that certain plants, to secure their services, produce seeds equipped with a special ball of fat called an elaiosome (from the Greek *ealion*, "fat," and *soma*, "body"), a very energizing, almost completely oleaginous structure that ants love. The exchange seems simple and at the same time very convenient for the plant: the ant takes the seed, carries it to the anthill, eats the elaiosome, and leaves the rest of the seed there, in a humid, sheltered place, rich in fertilizers, ideal for germination.

Ants are among plants' most wonderful partners; the communication systems and mutual assistance between these hymenoptera and plants continue to fascinate scientists. A fairly recent discovery has shed light on the services rendered by the *Camponotus* (a kind of ant which is also involved in certain plants' defense, and with which there seems to be an especially close relationship) to certain carnivorous species, among them in particular the nepenthes. We've discussed these plants and the slippery walls of their fearsome trap-sacs, which prevent their victims from getting out (see chapter 3).

The nepenthes trap animals by producing nectar around the sac and luring them inside. For the trap to work, however, the sac's walls must always stay clean, to keep them as slippery as possible: if detritus or dust accumulates, the animals will find a foothold up to safety. Hence the important alliance with *Camponotus* ants, which in return for a little nectar voluntarily turn up to keep the traps always clean. It seems that even the plant world's most terrible "death machines" need friends!

CHAPTER 5
Plant Intelligence

In biology, we call a species "dominant" when it obtains more living space at other species' expense, thus showing more adaptability to the environment than its competitors and a superior capacity to solve the problems faced by each living being in the struggle for survival. The more abundant a species is, the greater its specific gravity within the ecosystem.

For instance: what would we say if we discovered that a faraway planet is 99 percent inhabited by a certain life form? We'd say the planet is dominated by that life form. Now let's come back to Earth. What do we say about our planet? That it is dominated by humans. Now, are we really sure that this thought, so reassuring in many ways, corresponds to reality? On Earth, 99.7 percent of the biomass (estimates range from 99.5 to 99.9 percent, so we've averaged them), or the total mass of everything that is alive, isn't composed of humans, but of plants! The human species, together with all the other animals, represents a mere 0.3 percent.

Given this state of things, our planet is certainly green; Earth is an ecosystem inarguably dominated by plants. But how can that be: how could the stupidest and most passive beings on the planet have achieved this primacy? We've just said that obtaining greater space at the expense of other species indicates greater adaptability, that is, superior problem-solving ability. So why, of all living things (in terms of mass, remember, not in the number of species), do animals make up only 0.3 percent, and why, of that 0.3 percent, do humans make up an even smaller percentage? Or putting the question another way, how do we reconcile this fact with the completely human assumption that we are the dominant species, we can control the planet, and we have greater rights than other species? This subject would be much easier to deal with rationally if it had less influence in our collective consciousness and were a matter of ordinary (and neutral) scientific inquiry: On planet Earth, is there really only 0.3 percent animal life compared to 97.7 percent plant life? Then plants are the dominant beings, while there are only trace amounts of animals. There can be only one explanation: plants are much more advanced, adaptable, and intelligent beings than we're inclined to think.

Can We Speak of "Plant Intelligence"?

Why is the word *intelligence* so jarring when used to refer to the plant world? The question will be answered in the course of this chapter, but for now, let's recall that thousands of years of prejudice and false notions condition our thinking about plants and ways of referring to them. We'll review some of the themes we've discussed thus far in order to explain the many valid reasons for using the expression *plant intelligence*.

Unlike animals, plants are stationary beings and live anchored to the soil (though not all do). To be able to survive in this condition, they have evolved ways of feeding themselves, reproducing, and defending themselves differently from animals, and they have constructed their bodies modularly in order to cope with external attacks. Thanks to this structure, animal predation (for example, an herbivore eating part of the leaves or the stem) isn't a serious problem. A plant doesn't have individual organs such as a brain, a heart, lungs, one or more stomachs because, if it did, their injury or removal (by the herbivore just mentioned) would jeopardize the entire organism's survival. In plants, no single part is essential; and, in fact, the structure is mostly redundant, made up of repeated modules that interact with one another and that in certain conditions can even survive autonomously. These characteristics make plants very different from animals and more like a colony than an individual.

One consequence of their having a structure so different from ours is that plants seem very distant from us, alien, to the point that sometimes it's even hard for us to remember they're alive. The fact that we share with almost all animals a brain, a heart, one or more mouths, lungs, stomachs makes them seem close and comprehensible. But with plants, it's completely different. If they don't have a heart, does that mean they don't have circulation? If they don't have lungs, do they not breathe? If they have no mouth, do they not eat? And without a stomach, do they not digest? As we've already seen, there's a good plant answer to each of these questions, and all of the functions can be carried out even in the absence of individual organs to control or perform them. So now let's try asking ourselves: since plants don't have brains, can they not think?

The first prejudice about plant intelligence comes from just this doubt: how can a certain function be carried out without an organ designed for it? Yet we've already seen that plants eat without a mouth, breathe without lungs, see, taste, feel, communicate, move, despite lacking sensory organs like the ones we have. So why doubt that they can think? No one could deny that a plant feeds or breathes, so why is solely the hypothesis that they think adamantly rejected?

Here we need to step back and ask ourselves: what is intelligence? Because this concept is so broad and difficult to circumscribe, naturally there are many different definitions (the drollest being that "There seem to be almost as many definitions of *intelligence* as there [are] experts asked to define it," from psychologist Robert Sternberg).

So our first task is to choose the definition that fits our situation. For plants we could use a rather broad definition: "Intelligence is the ability to solve problems." There certainly are others that might work perfectly well, but let's stick with this one. An interesting alternative would be to view intelligence as a uniquely human endowment, because it is linked to abstract thought or to some other typically human cognitive ability, while all other beings have "abilities" of a different nature, which we should name appropriately. That sounds reasonable. But is it true? What are the irreplicable characteristics that make us human?

What Can We Learn From Artificial Intelligence?

The typical and inimitable characteristics of our intelligence aren't easy to pinpoint. For help, we can look to studies in artificial intelligence (AI), a field that has spent decades exploring what constitutes the essence of human intelligence, and what distinguishes it from its mechanical counterpart. To answer precisely these sorts of questions, the greatest experts in artificial intelligence in the world meet every year to compete for the Loebner Prize, in which computer programs perform what's known as "The Turing Test." The test is named after the great mathematician Alan Turing (1912–1954), one of the fathers of information science who, in 1950, wondered whether a time would come when machines could think; and if they could, how would we know?

Instead of pursuing complicated theoretical models or wading through definitions of intelligence, Turing proposed an experiment that seemed very simple: a panel of people, each of whom, by means of a computer terminal, would converse on any topic with two invisible interlocutors, one a software program, the other a human being. The judges' task was to decide which interlocutor was the human and which the machine.

Turing stipulated that a machine would be considered to have passed the test when it fooled 30 percent of the judges after five minutes of conversation, and that until this happened, the test should be repeated. He predicted this would happen by the year 2000, and "one will be able to speak of machines thinking without expecting to be contradicted."

To date, no machine has fooled 30 percent of the judges, but that moment of surrender is fast approaching, and we're nearing the point where software can simulate a human conversation perfectly. Will we then actually be able to speak of thinking machines? According to Turing, yes. What would change for us humans at that point? It's hard to say.

For thousands of years, we were sure that we were the most exalted of living beings, and at the center of the universe, but in recent times that conviction has been painfully contradicted, and our certainties profoundly shaken. Just think: first we had to abandon the geocentric system, recognizing that we live on a very insignificant planet in a galaxy at the edge of the universe. Then we had to accept our resemblance to other animals, and even our descent from certain animals. What a slap in the face!

At that point we started erecting impassable barriers, to distinguish ourselves from the rest of creation: humans are the only beings that use language (not true), syntactic rules (not true), tools (not true—even octopuses use them!). At one time, at least, we were the only ones that could perform complex mathematical calculations, but today nobody can compete with a calculator that you can buy for a few dollars. Over the course of a few centuries, we've been forced into gradual but inexorable retreat, a retreat with no end in sight and with several fundamental implications. For example: what is implied by the fact that machines are increasingly capable of imitating and surpassing some of the intellectual characteristics that we once thought were our exclusive domain? Today, a computer can defeat our best chess champions; flawlessly memorize almost limitless quantities of any kind of data; make predictions; translate; and even create music (though not great music). Generally, we respond to these successes of artificial intelligence by saying that none in itself demonstrates true intelligence. But if, going on this way, one day we were to see that everything we assumed was the exclusive property of our intellect can be replicated and even improved upon by a machine, shouldn't we acknowledge our lower position relative to it? In short, is it wiser to make intelligence a bulwark of our difference from other living beings (and intelligence isn't the only thing we use this way), or instead to admit that intelligence is something we have in common with all other species of animals and plants?

Intelligence Unites, It Doesn't Divide

We're not embarrassed to acknowledge the intelligence of many animals, because they've shown their capacity to obtain food by using tools, develop a language, get out of a labyrinth, or solve other types of problems. Now let's ask: can plants do the same? Yes, they do it all the time! They defend themselves from predators by using complex strategies that not infrequently involve other species, are assisted by trustworthy "transporters" in pollination, circumvent obstacles, help one another, can hunt or lure animals, move to reach food, light, oxygen. Then why not admit that plants fully deserve to be called intelligent beings? Instead of denying a fact that's plain to everyone who has really observed their behavior, we should consider their way of solving problems a source of precious information for us humans, too.

Intelligence is a property of life, something that even the humblest single-celled organism must possess. Every living being is continuously called upon to solve problems that essentially aren't so different from the problems we face. Think about it: food, water, shelter, companionship, defense, reproduction—aren't they the underlying factors in our knottiest problems? Without intelligence there can be no life. Accepting this plain truth shouldn't trouble us: the intelligence of human beings is obviously much greater than that of a bacterium or a unicellular alga. But the fundamental point is that this difference is only quantitative, not qualitative.

If we define intelligence as the capacity to respond to problems, then it's not possible to demarcate any kind of threshold above which intelligence appears and below which there are only automatons (that is, beings which respond to environmental stimuli in an automatic way). Anyone who disagrees, and still maintains that certain animals are intelligent and others not, should be willing to tell us at exactly what point in evolution intelligence appears.

Let's try it: Humans are intelligent, certainly no one would question that! And primates? Intelligent, too—it's been proven. Dogs? They sure are. Cats? Anybody who lives with one will swear to it. How about mice, aren't they smart? Of course! What would you say about ants? Certainly. Well, what about octopuses? Reptiles? Bees? And amoebas, which can get out of a maze or anticipate repetitive phenomena? Then is there a threshold above which intelligence magically appears? Or instead, in a way that's more evolutionarily accurate, should we think of intelligence as something that's natural to life? Besides, if it weren't, we would have even harder problems to solve.

If we hypothesize intelligence as being linked to the crossing of some sort of threshold, we have to ask if the threshold is fixed, and, therefore biological, or instead cultural, and thus variable according to time and place. In the 1800s, few people thought that an animal could be defined as intelligent. Today no scientist would dream of denying the intelligence of a monkey, a dog, or even a bird. A sizable literature even exists on bacterial intelligence. So why not speak of plant intelligence?

In fact, as we know well, every plant continuously registers a great number of environmental parameters (light, humidity, chemical gradients, the presence of other plants or animals, electromagnetic fields, gravity, and so on) and, on the basis of those data, has to make decisions regarding food, competition, defense, relations with other plants and animals—activity that's hard to imagine without resorting to the concept of intelligence! Moreover, one of the greatest geniuses who ever lived, Charles Darwin, realized over a century ago that plants possessed unaccountably evolved abilities. But the times were inhospitable and Darwin, already strenuously occupied with the defense of others of his theories, including the theory of the evolution of species, which would bring him immortal

Figure 5-1. Charles Darwin. The extraordinary botanist was an admirer of plants' abilities (drawing by Stefano Mancuso).

fame, limited himself to speculating on this subject in several works on botany, and especially in his "notebooks," whose extraordinary scientific significance has only recently become clear. Of the six books Darwin devoted to botany, one is essential for understanding what he really thought about plants. It is the only one full of experimental data, a revolutionary book even in its title: *The Power of Movement in Plants* (see chapter 1).

Charles Darwin and the Intelligence of Plants

Charles Darwin was introduced to the plant world while a theological student at Cambridge, attending classes taught by the botanist and geologist John Henslow (1796–1861). He soon became Henslow's inseparable student, known by other professors as "the man who walks with Henslow." Henslow's importance in Darwin's life was fundamental. It was on Henslow's recommendation that Captain Robert FitzRoy accepted Darwin as a "gentleman companion" aboard the HMS *Beagle*. And it was from Henslow that Darwin learned the basics of botany and, above all, a passion for the plant world that would last the rest of his life. Starting in those first years at Cambridge, and through the following decades, Darwin devoted himself rapturously to the study of plants, seeking evidence for the theory of evolution in these fascinating creatures and remaining interested in them almost until the day he died. (His last known letter, written just nine days before his death, was about a plant.)

The key to understanding the revolutionary import of *The Power of Movement in Plants*, the book that was destined to change the history of botany, is contained in its final paragraph, where Darwin, in what we know had become a customary practice for him, stated the essential conclusions of his research. On the relation between the movements of a plant's root system and the existence of a form of plant intelligence, he wrote:

> It is hardly an exaggeration to say that the tip of the radicle thus endowed [with sensitivity] and having the power of directing the movements of the adjoining parts, acts like the brain of one of the lower animals; the brain being seated within the anterior end of the body, receiving impressions from the sense-organs, and directing the several movements.

Moreover, in the 500-plus pages of his groundbreaking essay, the brilliant scientist describes plants' numerous movements, concentrating for more than three-quarters of the book on the movements of the roots. He focuses his observations on the roots precisely because this is the part of the plant where he sees the greatest number of similarities to animals' movements, and also the best examples of behavioral resemblances to other living beings. In fact, it's in the roots, or to be exact, in the root tip, the point of each root, that we find the typical sequence of phases that mark intelligence: perception of environmental stimuli, decision making about the direction of movement, purposeful movement.

Darwin was convinced that there wasn't such a substantial difference between a worm's brain, or the brain of any other lower animal, and the tip of a root:

> We believe that there is no structure in plants more wonderful, as far as its functions are concerned, than the tip of the radicle. If the tip be lightly pressed or burnt or cut, it transmits an influence to the upper adjoining part, causing it to bend away from the affected side; [...] If the tip perceives the air to be moister on one side than on the other, it likewise transmits an influence to the upper adjoining part, which bends towards the source of moisture. When the tip is excited by light [...] the adjoining part bends from the light; but when excited by gravitation the same part bends towards the centre of gravity.

Darwin was the first to notice that the root tip is a sophisticated sense organ capable of registering and reacting to different parameters. And after arguing that the root tip is sensitive to external stimuli, he further suggested that this was the area where signals were generated that could induce movement in adjacent parts of the root. In his experiments, he observed that after the tip is surgically removed, the root loses much of its sensitivity: for example, it can no longer perceive gravity or distinguish the density of the soil. Thus, Darwin formulated what a century later would become known as the "root-brain hypothesis," at the same time initiating the study of root physiology. It was an ineluctable decision, given what Darwin described as the root's "importance for the life of the plant."

As with many of Darwin's other ideas, the reception by the scientific community was far from enthusiastic. The greatest opposition came from German botanists—just as he had predicted. In a letter to Professor Julius Victor Carus in 1879, he wrote, "Together with my son Francis, I am preparing a rather large volume on the general movements of plants, and I think that we have made out a good many new points and views. I fear that our views will meet a good deal of opposition in Germany. . . ."

What motivated their hostility was not solid scientific thinking but, more than anything else, the resentment felt by the great botanist Julius von Sachs (1832–1897) at what he took to be Darwin's unjustified invasion of territory. Sachs, at that time, was a highly respected botanist, and he considered Darwin's work to be the findings of an amateur—"a country house experimenter"—whose research could not be compared to his own serious work in plant physiology.

Following the publication of *The Power of Movement in Plants*, Sachs asked one of his assistants, Emil Detlefsen, to repeat Darwin's experiments, particularly those concerned with the behavior of the root after removal of the cap (the outer part of the root tip). His obvious goal was to refute the validity of Darwin's conclusions. Detlefson dutifully set about repeating the experiments, but his work, as was later ascertained, was carelessly executed because of the low regard for Darwin in Sachs's laboratory, and produced different results from Darwin's.

Sachs's response at that point was vehement. He accused both Darwins, father and son, of having done the experiments improperly (exactly like "amateurs") and of having jumped to false conclusions. They, of course, defended their work.

The clash between these celebrated botanists reverberated in the scientific community, prompting Sachs's former student Wilhelm Pfeffer (1845–1920), a renowned botanist in his own right, to repeat the experiments soon after. His motivation was genuine scientific spirit, and his research produced results identical to those obtained by the Darwins! Pfeffer unhesitatingly acknowledged the greatness of the two scientists in his *Lehrbuch der Pflanzenphysiologie (Manual of Plant Physiology*, published in 1874), a book that the ever more bitter Sachs dismissed as "nothing more than a pile of undigested facts."

Today, of course, we know that Darwin was right. In fact, the root tip is even more advanced than Darwin imagined, able to detect numerous physicochemicial parameters in the environment.

The Intelligent Plant

We'll begin this section by restating the obvious: plants don't have a brain. We've said this several times in the preceding pages, but we repeat it here to be even more explicit: plants have no organ that remotely resembles the brain as we understand it.

In humans, the brain is the seat of intelligence, and indeed we use expressions like "has a brain" or "is brainless" to refer to the presence or absence of evident intellectual ability in a person.

Like most of the familiar animals we recognize as having some form of intelligence, we're endowed with this extraordinary organ whose complexity and functioning for the most part still elude us, and without which, at least among members of the animal kingdom, there is no cognition. Now the first question to ask is this: Is the brain really the only production site for "intelligence"? Would a brain without a body still be intelligent, or on the contrary, would it appear to be only a group of cells without any special characteristics? Would we be able to find in it any traces of intelligence? The answer, unequivocally, is no. The brain of our greatest geniuses is no more intelligent in and of itself than the stomach. It's not a magic organ and it certainly couldn't create anything on its own. Information coming from the rest of the body is fundamental for any intelligent response.

Well, in plants, cognition and bodily functions are not separate but are present in every cell: a real, living example of what artificial intelligence scientists call an "embodied agent," that is, an intelligent agent that interacts with the world with its own physical body.

We've emphasized repeatedly that evolution has given plants a modular structure, not concentrating functions in individual organs but distributing them in the entire living

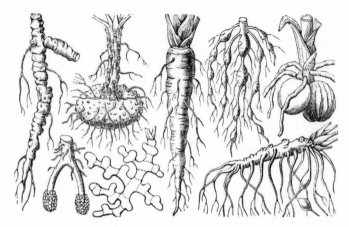

Figure 5-2. Examples of root systems. The roots are the hidden half of plants, and the most interesting. The illustration shows various types.

being. This was a fundamental strategic choice, as we've seen, allowing plants to lose even substantial parts of their own organism without risk to their survival. So a plant has no lungs, or liver, or stomach, or pancreas, or kidneys. And yet it can still carry out all the functions performed by these organs in animals. Then why must the lack of a brain prevent it from being intelligent?

Let's look at the root, the part of the plant in which, as we've seen, Darwin perceived decision-making and directing capacity. Its point, the root tip, has the universally acknowledged function of directing underground growth and soil exploration in search of water, oxygen, and nutritive substances. Now, it would certainly be easy to hypothesize an automatic growth, guided by simple instructions such as "Look for water" or "Grow downward." In that case, there really would be nothing to the root's job: detect water and develop in that direction, or grow downward, guided by the force of gravity. But in fact, the root has a much more complex function than this. It has many tasks to perform and different needs to balance; the tip is required to make complex assessments as it directs the root in its exploration of the soil.

Oxygen, mineral salts, water, and nutrients are usually found in different areas of the soil, sometimes far apart. So the root must continually make crucial decisions: grow toward the right and reach the phosphorus it needs so much, or toward the left and reach nitrogen, always in short supply? Grow downward in search of water, or upward where there's more likely to be good air to breathe? How to reconcile needs that call for opposite choices? In addition, let's not forget that on its path the root often encounters obstacles to avoid, or outright enemies (another plant, parasites) which it must "dodge" and defend itself against. And this is only the start, because then the local needs of an individual root must be reckoned against the global needs of the entire plant, which may be different.

So many variables, and each one fundamentally important for life! How can a plant keep all its roots from growing toward the same point, perhaps in search of water? This would be a real, concrete danger if root growth were controlled automatically. To address this question, at this point it's important to understand how the fabulous root tip is made and how it functions.

The tip is the extreme point of the root, and its size varies from species to species, from a few tenths of a millimeter (for example in *Aradiopsis thaliana*) to a couple of millimeters (as in corn). Usually white in color, the tip is the vital part of the root, the part that reaches out and has the greatest sensory capacities; it is also an area of intense electrical activity based on action potentials, electrical signals closely resembling the signals occurring in the neurons of animals' brains. Each plant has millions of root tips: the root system of even a very small plant may have more than 15 million!

Each root tip continuously detects numerous parameters such as gravity, temperature, humidity, electric field, light, pressure, chemical gradients, the presence of toxic substances (poisons, heavy metals), sound vibrations, the presence or absence of oxygen and carbon dioxide. A stupendous list, but far from exhaustive: the number

of these parameters is constantly being updated by researchers and increases from year to year. The root tip registers these continuously and guides the root on the basis of a real calculus that takes into account the plant organism's different local and global needs.

No automatic response could possibly answer the requirements of the root tip! In fact, each tip is a true "data processing center"; it doesn't work alone, but in a network with millions of others that make up the plant's root system.

Figure 5-3. The root tip. Every root tip is a sophisticated sensory organ.

Each Plant Is a Living Internet Network

Until now, we've been discussing the functioning of an individual root tip, but even small plants such as rye or oats may have tens of millions of root tips, while in a tree—though no studies exist of this topic specifically—it's reasonable to assume there are hundreds of millions. How do all these roots work together? The individual root tips of the same plant shouldn't be considered independently, but as a network capable of functioning collectively.

To understand what we're talking about here, let's think about the Internet, the largest and most powerful network ever created by human beings.

For the solution of very complex calculations, research in recent decades has developed in two different directions that closely relate to our discussion about plants. On one hand, it has led to the production of individual megacomputers, more and more powerful and capable of carrying out astounding quantities of calculations in very short times (the IBM computer Sequoia, in operation since 2012, in one hour can complete a series of calculations that would take 6.7 billion people working with hand calculators 24 hours a day 320 years to perform); on the other hand, it's been directed toward utilizing the immense calculating capacity possessed in total by a network such as the Internet. These two opposite strategies recall those played out by evolution to increase the calculating capacity of living organisms: on one hand, increasingly large and better-performing individual brains (with humans in the Sequoia's role); on the other, a distributed intelligence, such as we see in insect societies and in plants.

The calculating speed (in units of time) of a supercomputer is fundamental and will always exceed that of a computer network like the Internet, but the security offered by the network is also an important factor that should not be underestimated. The Internet network, the first version of which (Arpanet) was created by DARPA (Defense Advanced Research Projects Agency, a program of the U.S. Department of Defense), was initially conceived and constructed modularly in order to be able to withstand a large-scale nuclear attack. Even if most of the individual computers constituting the network were destroyed—this was the crux—the network's modular construction would ensure its survival and thus the transmission of data.

Does that sound familiar? It's the same strategy adopted by plants: millions of root tips working in a network so that the destruction or predation even of an important part doesn't compromise the network's survival. By itself, one root tip doesn't have great calculating capacity, but together with the other root tips it is capable of extraordinary feats, just like an ant, which cannot map out strategies on its own, but by working with other ants creates a society that's among the most complex and structured in nature.

But how do the roots manage to collaborate and coordinate with each other? We don't yet know for certain, but recent research allows us to make interesting hypotheses.

The root system is first of all a physical network in which the roots are connected to each other anatomically. And yet, this connection doesn't seem at all to be the most important. In fact, the signals enabling each root to communicate with

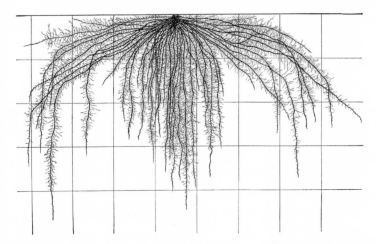

Figure 5-4. Root system of an eight-week-old corn plant. A root system is made up of tens of millions of root tips.

the others probably don't travel inside the plant. How is this possible?

Returning to our ant comparison, try imagining the root tips as a colony of insects: the ants are not even physically connected to each other, and yet they act in a coordinated way through chemical signals. Maybe the roots do likewise? Plants are true masters of the art of producing chemical molecules of every kind and for every purpose; so it wouldn't be surprising if the hypogeal (underground) parts, like the aerial parts, released chemical signals to communicate with one another.

But now we've entered the realm of hypothesis, and so we should consider other possibilities. For example, the root tips may be extremely sensitive to electromagnetic fields, including those caused by nearby root tips, and may behave accordingly. Or they may perceive sounds made by other roots as they grow. As we know (see chapter 3), recent studies have shown that each growing root gives off sounds similar to "clicks," which may be perceived by the root tips of plants nearby. In that event, there would be a very convenient communication system: as we've seen, these sounds seem not to be produced intentionally by the plant, but rather to result from the breaking of the cell walls during growth. In that case, this would be a so-called parsimonious signal, a signal that achieves its purpose while sparing the plant the effort, or rather the energy expenditure, necessary to produce it.

A Swarm of Roots

Picture one of those fast-moving black clouds that form in the evening skies in spring, a flock of thousands of birds flying together, making evocative images. Until the 1970s, their coordinated movement was a true mystery; in theory, flying so close together they should have continually collided. Scientists fumbled around in the dark for answers, with some even suggesting (in serious scientific journals) that the birds were endowed with . . . telepathy! Actually the explanation is quite simple, but the veil on this mystery wasn't lifted until recently.

In a flock, each bird follows a few basic rules, such as keeping a certain number of centimeters' distance from the birds in front of it and to its right; this suffices to ensure the coordinated flight of all the birds, even if there were thousands of them performing daredevil maneuvers! A system at once both elementary and functional, which would not likely have evolved only in birds' flight. Indeed, one of the most widely accepted theories on the roots' functioning suggests that they may behave in the same way as a swarm.

According to this theory, each root tip maintains a preset distance from the root tips around it. This behavior in and of itself enables coordinated growth and thus also the best possible exploration of the soil, without any need of a higher type of volition; that is, a brain directing each individual root tip's operation. Lacking a specific organ to supervise cognitive functions, plants developed a form of distributed intelligence, typical of swarms and many other living beings: when the individuals constituting a swarm are together, they display so-called emergent behaviors, which don't exist in individual organisms.

In recent years, this phenomenon has been observed and studied systematically, with exciting results. Even human beings have been shown to activate some of the dynamics of emergent behavior when together in a group. The classic example is thousands of people clapping in a theater: recent studies have shown how the applause at first is asynchronous (each person starts clapping his or her hands independently of everyone else), but after a few seconds tends to synchronize until a uniform sound results. Naturally, this synchrony is involuntary and is the expression of an emergent behavior. An observer might wonder: how do thousands of people manage to clap their hands in a coordinated way? Who decides the rhythm? And who tells the others the rhythm to follow?

Models of emergent behavior have been used to describe many human activities, from the ability to walk on very crowded sidewalks without treading on each other, to stock market trends. Think about it: the stock market tells us the worth of businesses all over the world, effectively dominates politics, and has considerable influence over our individual fates, all without any central control. Indeed, there's no entity dedicated to overseeing its overall functioning: investors know only a very limited number of companies within their portfolios, and simply follow market rules. Ultimately, the behavior of the stock exchange derives solely from the interactions of individual investors. Like the tips of a root system or ants in a swarm, they amount to nothing by themselves, but together develop incredible capacities.

The similarities between plants and animals involve this type of behavior, too, but with a significant difference. In the animal world, swarms are formed by great numbers of people, mammals, insects, or birds. But in plants, these dynamics actually come into play inside one plant, between its roots. In short, every single plant is a swarm!

The Aliens Are Here (Plant Intelligence as a Model for Understanding Extraterrestrial Intelligence)

The study of plant intelligence points up a very interesting aspect of research on intelligence in general: how difficult it is for us humans to understand living systems that think differently from us. Indeed, we only seem able to appreciate intelligences that are very similar to ours.

The same kind of problems surface when intelligence is spoken of in reference to organisms without a brain, such as—excluding plants for the moment—bacteria, protozoa, and molds. Although some (bacteria and protozoa) are so simple as to be composed of only one cell, they too nevertheless display behavior that—if their size were more impressive and, above all, if they had a brain—we wouldn't hesitate to term intelligent: amoebas solve mazes, while molds can map out a territory more efficiently than any software invented by human beings. However, in these organisms as in plants, our brain bias leads us to deny the existence of any sort of thinking capacity—an attitude that seems based more on traditions and preconceptions than on scientific reasoning. And yet the study of plant intelligence could turn out to be fundamental for the progress of humanity; in fact, it would enable us to look at our own mind with different eyes.

Suppose that one day we came into contact with an intelligent alien life form: would we be able to, if not communicate with it, at least recognize it? Probably not. It seems that we humans, unable to conceive of intelligences different from ours, rather than searching for alien intelligences, are in continual search of our own intelligence, lost somewhere in space. If alien forms of intelligence should really exist, they would have evolved in organisms very different from us. Their chemistry would be different from ours and they would inhabit environments totally unlike those that we know.

How could we ever hope to recognize them when we can't discern the intelligence of plants, organisms with which we share a great part of our evolutionary history, the same cellular structure, the same environment, the same needs? Let's ask ourselves, for example: Why should an intelligence that evolved on another planet and under completely different conditions from ours use the same means of communication we use, based on wave phenomena? The voice, sound, and radio and television communications are all, in fact, based on wave propagation. Other living things, plants among them, use different systems to communicate, some based on the production of chemical molecules. They are extremely efficient methods, very well suited to information transmission, but we still know very little about them, although they're used by numerous species on our own planet!

All that was necessary to make plant intelligence utterly alien to us was that they were slower than us and lacked individual organs similar to ours; imagine if they'd been born and evolved light-years away. And yet, precisely because they're physically and genetically different from but fundamentally so close to us, plant organisms could be an important model for the study of intelligence, and could help us rethink our approaches and tools in searching for alien intelligence in space.

Plants' Sleep

Sleep remains one of science's great mysteries, though thousands of philosophers and researchers have inquired into its nature. Aristotle was among the first to speculate on this subject:

> With regard to sleep and waking, we must consider what they are; whether they are peculiar to soul or to body, or common to both; and if common, to what part of soul or body they appertain: further, from what cause they are attributes of animals, and

whether all animals share in them both, or some partake of the one only, others of the other only, or some
partake of neither and some of both.

Two thousand years later, many of these questions are still
unanswered.

What is the purpose of sleep? What is the nature of dreams,
and how do they function? Before Aristotle, the Greek philosopher Heraclitus of Ephesus (c. 535–c. 475 BCE) had
said: "Man kindles a light for himself in the nighttime," a
view that would be clarified and validated by psychoanalysis,
according to which dreams reveal parts of our unconscious.
Today, we know that sleep affects the processes of learning
and memory and thus acts upon the noblest functions of the
brain. For centuries, science believed that only humans and
a few higher animals had the capacity to sleep, but recently
insects have joined this select group. The discovery in 2000
that even *Drosophila melanogaster*, the common fruit fly, goes
to sleep stirred a revolution in the study of sleep in animals.
If even the simplest animal is capable of sleeping, sleep must
be acknowledged as one of the essential components of life!

And plants? Do they sleep, too? This only seems to be an
idle question, and in recent years it has interested a growing
number of scientists. Specifically, if plants are endowed with
intelligence and the ability to think, sleep could be an activity
related to these properties.

As we've mentioned in chapter 1, *Somnus Plantarum* is the
title of a little-known treatise written in 1755 by Carl Linnaeus,
culminating his studies of the different positions assumed by

certain plants' leaves and branches during the night. François Boissier de Sauvages de Lacroix (1706–1767), a celebrated botanist from Montpelier, had sent Linnaeus as a gift a specimen of *Lotus corniculatus*, whose flower he wanted to study. The delicate plant, transported from the Mediterranean coast to the cold of Uppsala, took several months to adapt to the new climatic conditions, but one May morning, in the greenhouse and after constant tending, it finally bloomed. Linnaeus observed this first morning blossoming and went back to see the plant the same day in the late afternoon. To his astonishment, the delicate yellow flowers he'd admired only a few hours before were no longer there. What had happened to them? The next morning when he returned to observe the plant, he found them in place, perfectly fresh. The mystery was soon solved: the phenomenon Linnaeus had witnessed was a typical example of what modern botanists call "nictinasty" (from the Greek *nux*, "night," and *nastos*, "compact"), the capacity many plants have to change their leaves' and flowers' positions from day to night. In the case of *Lotus corniculatus*, Linnaeus noticed that near dusk, the lotus extended and lifted its leaves and brought them together around each group of flowers, which then became invisible to even the most observant eyes. At the same time, the peduncles slightly drooped and the pedicels bent toward the ground. Thus began Linnaeus's interest in the so-called sleep of plants, which led him to plan a "flower clock"—a garden in which one could tell the time simply by observing the plants' behavior.

Actually, the first observations concerning the circadian movements of plants occurred long before Linnaeus's time, in

Figure 5-5. Leaves in diurnal and nocturnal phases. From top left: *Desmodium gyrans (Codariocalyx motorius), Lotus creticus, Cassia pubescens, Cassia corymbosa, Nicotiana glauca, Marsilea quadrifoliata.*

ancient Greece. In the fourth century BCE, Androsthenes, the scribe of Alexander the Great, noticed that the leaves of the tamarind were open during the day and closed at night. Similar observations are found frequently in the writings of botanists, in different times and places. In 1260, Albertus Magnus (c. 1193–1280) in *De vegetalibus et plantis* ("On Vegetables and Plants") described the periodic daily movements of the pinnate leaves of certain legumes, while in 1686, John Ray (1627–1705) in *Historia Plantarum (History of Plants)* first spoke of the "phytodynamic" phenomena of plants between day and night.

In 1729, Jean-Jacques d'Ortous de Mairan (1678–1771), studying mimosa plants that open and close their leaves approximately every twenty-four hours, concluded that they must have a kind of internal clock controlling the leaves' movement. So sleep had been observed in plants at various times before Linnaeus, but Linnaeus deserves the credit for treating the subject systematically. Linnaeus gave no explanation for this behavior in plants, though he had guessed that it was mainly light, not temperature, that caused the leaves' movement. He limited himself instead to classifying all the plants exhibiting this phenomenon and to naming their nighttime position the "sleep of plants."

In contrast to attempts made in more recent years, Linnaeus didn't treat plants' sleep in a metaphorical way, but rather viewed this behavior as a phenomenon wholly analogous to the sleep of animals. For example, plants change position during the night. This movement isn't easy to discern in species with coriaceous leaves, such as the oak, the olive tree, and the laurel, but it's clearly visible in all species with more-delicate leaves. Like animals, plants assume a position for their night's rest that differs from species to species. Just as the duck hides its head under its wing, the ox rests on its side, and the hedgehog curls into a ball, so the spinach plant's leaves straighten up toward the top of the stem, and the impatiens and the bean point their leaves downward; trefoils, like the *Lotus corniculatus* studied by Linnaeus, bring their leaves together around the flowers, while their relatives, the lupines, turn their leaves downward; the oxalides, whose leaf is composed of three heart-shaped leaflets, fold them in half along the midrib and leave them hanging upside down from the far end of the petiole. This multiplicity of nocturnal positions follows a general law: the leaves, in fact, show a common tendency to assume the same position at night that they had during germination. So in one plant, the leaf rolls into the shape of a cylinder; in another, it folds like a fan; in a third, it closes in two along the midrib; but in general, each tends to position itself during the night in the same way as in its first stages of growth.

The similarities with animals don't stop here, however. In the plant world, too, for example, the proclivity for sleep is greater in youth, while as the plant ages, the waking state lengthens and going to sleep becomes somewhat difficult: in this respect, the plant's behavior is completely analogous to that of animals (and humans!). There comes a time for some plants when the propensity for sleep diminishes and the leaves respond less and less to the triggers that cause them to assume their nighttime positions. But for what purpose do the leaves open during the day and close at night? And what events trigger sleep and waking in plants? There are no answers yet to these questions, but as research continues, plants could be used as models for the study of sleep, providing scientists with a genetic tool with which to investigate the mechanisms and disorders of this important biological function.

Conclusion

Two attributes come to mind instinctively when we think of plants: immobility and insentience. These aren't just any qualities, but very particular ones, and they largely determine our estimation of the plant world. Yet in contrast to what we've thought for hundreds of years, they are not innate properties of plants, but a simple and enduring cultural construct that originated with Aristotle. In his conception, the plant world is "a level below" that of animals, being devoid of *anima*, a concept which for him meant "motor principle," directly linking it to the principle of movement. The capacity for self-motion distinguishes the living from the nonliving: so plants, which move very little, are at the borderline between life and nonlife.

The idea of plants' complete difference from animals began to lose sway only at the end of the nineteenth century, and it still is quite prevalent. Today, however, at least on the scientific level, it's clear that the difference between plants and animals is not qualitative but quantitative. Animals use the matter and energy produced by plants. Plants, in turn, use the sun's energy to fulfill their own needs. Thus animals depend on plants, plants on the sun.

This brings us to a more general conception of plant life and to an understanding of its role in the biosphere: plants are the mediators between the sun and the animal world. They—or rather their most typical cellular organelles, the chloroplasts—are the link connecting the activities of the whole organic world (that is, of everything we call life) with our solar system's energy center. Thus plants have a universal function for life on our planet. Animals don't.

The most recent studies of the plant world have demonstrated that plants are sentient (and thus are endowed with senses), that they communicate (with each other and with animals), sleep, remember, and can even manipulate other species. For all intents and purposes, they can be described as intelligent. The roots constitute a continuously advancing front line, with innumerable command centers, so that the whole root system guides the plant like a kind of collective brain—or rather a distributed intelligence—which, as the plant grows and develops, acquires information important to its nutrition and survival.

Recent developments in plant biology enable us to study plants as organisms with a proven capacity for acquiring, storing, sharing, processing, and utilizing information collected from their environment. How these brilliant creatures get information and process it in a way that results in consistent behavior is the focus of plant neurobiology.

Researchers studying plants' systems of communication and socialization now see on the horizon the development of new and previously unimagined technological applications. For some time now, there's been talk of plant-inspired robots, a real generation of plantoids destined soon to follow human-inspired robots (the so-called androids) and animal-inspired robots in the robotic evolutionary chain. Plans are also under way for the construction of plant-based networks, with the capacity to use plants as ecological switchboards and make available on the Internet in real time the parameters that are continuously monitored by the roots and leaves: we've called this type of network "Greenternet." Soon the plant Internet may become part of everyday life for all of us—we'll be able to get advance warnings of an approaching toxic cloud, information about the quality of our air and soil, news of impending earthquakes and avalanches. And now in the works is the design of *phytocomputers*—computers that use new algorithms based on the capacities and calculating systems of plants ("unconventional computing").

Beyond being a source of inspiration for robotics and information science, the plant kingdom may offer numerous innovative solutions to many of our most common technological problems. *Bioinspiration*, or finding impetus in the living world for devising new technological applications, originated centuries ago (think, for example, of Leonardo's studies of flying machines, inspired by bird flight). Our eyes have long been fixed on the animal kingdom (the realm closest to us) and have only recently begun to discern the treasure hidden in the plant realm, where one day we may find—among other things—the cure for many of humanity's most serious illnesses, new forms of clean energy, opportunities for innovation in man-made materials, and incalculable unexplored possibilities in the chemical and biological world.

Clearly the plant kingdom is not only an essential ingredient of life on our planet, but a great gift to human beings and to our intelligence—a gift we often heedlessly throw away. It's been estimated that human beings know scarcely 5 to 10 percent of the plant species present on Earth, and that from them we derive 95 percent of all our most important medicines. Each year, thousands of species we know nothing about become extinct, and untold gifts to humanity are lost with them. Perhaps knowing that plants perceive, communicate, remember, learn, and solve problems will help us someday to see them as closer to us and will also offer us the opportunity to study and protect them more effectively.

Considering the scientific evidence accumulated in recent decades, it was unsurprising when the Federal Ethics Committee on Non-Human Biotechnology (ECNH), established in 1998 by the Swiss Federal Assembly, released a document entitled *The Dignity of Living Beings with Regard to Plants: Moral Consideration of Plants for Their Own Sake* at the end of 2008. Though it may seem a stretch to invoke for plants a concept that has marked human history, the reference to plants' dignity can be understood as a first step toward legitimizing their rights, independent of human interests. It signifies that plants should be respected and that we humans have responsibilities in relation to them. If we regard these creatures as mere things, passive machines performing a program by rote, if we see them as insignificant except to the extent that they satisfy our interests and needs, then an attribute like dignity seems nonsensical. But if plants are active, adaptable, actually capable of subjective perceptions, and, above all, if they possess life in way that is totally independent of us, then there are excellent reasons to agree that the concept of dignity applies to them, too.

At the beginning of the twentieth century, Jagadish Chandra Bose (1858–1937), one of the first modern Indian scientists and a legendary figure in modern Indian history, a proponent of the fundamental identity between plants and animals, wrote: ". . . These trees have a life like ours. . . . They eat and grow . . . face poverty, sorrows, and suffering. This poverty may induce them to steal and rob, [but] they also help each other, develop friendships, sacrifice their lives for their children."

Many issues are still controversial, and much is yet to be discovered. But on this the Swiss Bioethics Committee— moral philosophers, molecular biologists, naturalists, and ecologists—unanimously agreed: plants cannot be treated arbitrarily. Their indiscriminate destruction is morally unjustifiable.

It must be pointed out that recognizing that plants have rights doesn't necessarily mean reducing or constraining their use. Just as recognizing the dignity of animals hasn't meant eliminating them from the food chain or banning experimentation altogether.

For centuries, animals, too, were considered unthinking machines. It is only in the past several decades that we've begun to guarantee them rights, dignity, and respect: animals are not things anymore. This change in perspective has led nearly all the most advanced nations to enact regulations designed to protect and defend animals' dignity. Nothing like this exists for plants. The discussion of their rights is only beginning, but it can't be put off any longer.

NOTES

Chapter 1

On sleep in plants (a subject discussed at greater length in chapter 5), see:

> —Aristotle. "On Sleep," "On Dreams," and "On Divination in Sleep." Translated by J. I. Beare. In Vol. 1 of *The Complete Works of Aristotle*. Bollingen Series, revised Oxford translation, edited by Jonathan Barnes. Princeton, NJ: Princeton University Press, 1984.
>
> —Linnaeus, C. *Somnus Plantarum*. Upsala, Sweden: 1755.

For the history of the idea that plants are like upside-down humans, see:

> —Repici, L. *Uomini Capovolti: Le Piante nel Pensiero dei Greci*. Bari: Editori Laterza, 2000.

The idea that plants were essentially immobile or that all their movements were involuntary was completely discarded thanks to the work of Charles and Francis Darwin. Their book was a true milestone in plant neurobiology; see:

—Darwin, C., and F. Darwin. *The Power of Movement in Plants*. London: John Murray, 1880. Reprint, Cambridge, UK: Cambridge University Press, 2009.

Francis Darwin's speech on the intelligence of plants is available in *Science*:

—Darwin, F. "The Address of the President of the British Association for the Advancement of Science." *Science* 18 (September 1908): 353–62.

Chapter 2

The theme of the sudden disappearance of human beings has been engrossingly explored by Alan Weisman, who imagined the behavior of other species after our extinction:

—Weisman, A. *The World Without Us*. New York: Thomas Dunne Books, 2007. www.worldwithoutus.com.

To date, there have been few comprehensive studies of the beneficial effects of plants with respect to stress, rehabilitation, attention, and diverse other psychophysical parameters; but see the following articles:

—Dunnet, N., and M. Qasim. "Perceived Benefits to Human Well-Being of Urban Gardens." *HortTechnology* 10 (2000): 40–45.

—Honeyman, M. K. "Vegetation and Stress: A Comparison Study of Varying Amounts of Vegetation in Countryside and Urban Scenes." In *The Role of Horticulture in Human Well-Being and Social Development: A National Symposium*, 143–45. Portland, OR: Timber Press, 1991.

—Tennessen, C. M., and B. Camprich. "Views to Nature: Effects on Attention." *Journal of Environmental Psychology* 15 (1995): 77–85.

—Ulrich, R. S. "View through a Window May Influence Recovery from Surgery." *Science* 224, no. 4647 (1984): 420–21.

—Mancuso, S., S. Rizzitelli, and E. Azzarello, "Influence of

Green Vegetation on Children's Capacity of Attention: A Case Study in Florence, Italy." *Advances in Horticultural Science* 20 (2006): 220–23.

Chapter 3

For an introduction to the world of carnivorous plants, see:

—D'Amato, P. *The Savage Garden*. Berkeley, CA: Ten Speed Press, 1998.

On the extraordinary world of the nepenthes, see the following:

—Clarke, C. *Nepenthes of Borneo*. Kota Kinabalu, Sabah, Malaysia: Natural History Publications, 1997.

———, *Nepenthes of Sumatra and Peninsular Malaysia*. Kota Kinabalu, Sabah, Malaysia: Natural History Publications, 2001.

Indispensable reading is Charles Darwin's *Insectivorous Plants*, originally published by John Murray (London, 1875); the book is now available in digital form from Darwin Online, edited by John van Wyhe, http://darwin-online.org.uk/.

For the first published description of the dionaea, see:

—Ellis, J. "Botanical Description of a New Sensitive Plant, Called *Dionoea muscipula*, or, Venus's Fly-trap, in a Letter to Sir Charles Linnaeus." In *Directions for Bringing over Seeds and Plants from the East-Indies and Other Distant Countries,* 35–41. London: L. Davis, 1770.

The digitized book is available online as a PDF from the Hunt Institute for Botanical Documentation at: http://huntbot.andrew.cmu.edu/HIBD/Departments/Library/Ellis.shtml.

On "protocarnivores," see this illuminating article:

—Chase, M., et al. "Murderous Plants: Victorian Gothic, Darwin and Modern Insights into Vegetable Carnivory." *Botanical Journal of the Linnean Society* 161 (2009): 329–56.

On the ability of plants to make sounds, see:
> —Gagliano, M., S. Mancuso, and D. Robert. "Towards Understanding Plant Bioacoustics." *Trends in Plant Science* 17, no. 6 (2012): 323–25.

On swarming behavior in plants, see:
> —Ciszak, M., et al. "Swarming Behavior in the Plant Roots." *PLoS ONE* 7, no. 1 (2012). doi: 10.1371/ journal. pone.0029759.

The discovery of carnivorous plants capable of capturing animals in the soil with special underground leaves is very recent. Thus there are still relatively few sources; the first article on the subject is:
> —Pereira, C. G., et al. "Underground Leaves of Philcoxia Trap and Digest Nematodes." *PNAS (Proceedings of the National Academy of Science of the United States of America)* (2012). www.pnas.org/content/early/2012/01/04/1114199109.abstract.

On Gottlieb Haberlandt's theory of "ocelli," see:
> —Haberlandt, G. *Sinnesorgane im Pflanzenreich zur Perception mechanischer Reize.* Leipzig: Engelmann, 1901.

The German text, no longer under copyright, can be downloaded online here: http://archive.org/details/sinnesorganeimp00habegoog.

Chapter 4

On the opening and closing of the stomata, see:
> —Peak, D., et al. "Evidence for Complex, Collective Dynamics and Emergent, Distributed Computation in Plants." *PNAS (Proceedings of the National Academy of Sciences of the United States of America)* 101, no. 4 (2004): 918–22.

On communication between plants, in particular regarding the roots' ability to distinguish relatives from non-relatives and the resulting behaviors of the plant, see the following:

   —Dudley, S., and A. L. File. "Kin Recognition in an Annual
   Plant." *Biology Letters* 3 (2007): 435–38.
   —Callaway, R. M., and B. E. Mahall. "Family Roots."
   Nature 448 (2007): 145–47.

On crown shyness, and for a modern and unprejudiced view of plants,
see Francis Hallé's fundamental book:

   —Hallé, F. *Plaidoyer pour l'arbre*. Arles, France: Actes Sud,
   2005.

The symbiotic origin of mitochondria and their importance in the
evolution of higher forms of life are discussed in the following articles:

   —Lane, N., and W. Martin. "The Energetics of Genome
   Complexity." *Nature* 467 (2010): 929–34.
   —Thrash, Cameron J., et al. "Phylogenomic
   Evidence for a Common Ancestor of Mitochondria and the
   SAR11 Clade." *Scientific Reports* 1 (2011): 13. doi: 10.1038/
   srep00013.

On plants' defensive strategy of enlisting the aid of natural enemies of
herbivorous insects, see the following article:

   —Dicke, M., et al. "Jasmonic Acid and Herbivory Differen-
   tially Induce Carnivore-Attracting Plant Volatiles in Lima Bean
   Plants." *Journal of Chemical Ecology* 25 (1999): 1907–22.

For research on round leaves capable of attracting bat pollinators, see:

   —Simon, R., et al. "Floral Acoustics: Conspicuous Echoes
   of a Dish-Shaped Leaf Attract Bat Pollinators." *Science* 333, no.
   6042 (2011): 631–33. doi: 10.1126/science.1204210.
Here is the abstract of this article:

   The visual splendor of many diurnal flowers serves to attract
   visually guided pollinators such as bees and birds, but it
   remains to be seen whether bat-pollinated flowers have

evolved analogous echo-acoustic signals to lure their echo-locating pollinators. Here, we demonstrate how an unusual dish-shaped leaf displayed above the inflorescences of the vine *Marcgravia evenia* attracts bat pollinators. Specifically, this leaf's echoes fulfilled requirements for an effective beacon, that is, they were strong, multidirectional, and had a recognizable invariant echo signature. In behavioral experiments, presence of the leaves halved foraging time for flower-visiting bats.

For the history of the diabrotica and the loss of the gene for caryophyllene production in modern American varieties of corn, see the following articles:

—Rasmann, S., et al. "Recruitment of Entomopathogenic Nematodes by Insect-Damaged Maize Roots." *Nature* 434 (2005): 732–37.

—Schnee, C., et al. "A Maize Terpene Synthase Contributes to a Volatile Defense Signal That Attracts Natural Enemies of Maize Herbivores." *PNAS (Proceedings of the National Academy of Sciences of the United States of America)* 103 (2006): 1129–34.

On the genetic modification necessary to reintroduce in new varieties of corn the original system of defense against nematodes, which was lost over the course of time through the selection of new varieties, see:

—Degenhardt, J., et al. "Restoring a Maize Root Signal That Attracts Insect-Killing Nematodes to Control a Major Pest." *PNAS (Proceedings of the National Academy of Sciences of the United States of America)* 106 (2009): 13213–18.

The idea that plants are capable of manipulating all animals and that they've succeeded even with humans is set forth and boldly documented by Michael Pollan:

—Pollan, M. *The Botany of Desire: A Plant's-Eye View of the World.* New York: Random House, 2001.

On the dispersal of seed by fish, see:

—Anderson, J. T., et al. "Extremely Long-Distance Seed Dispersal by an Overfished Amazonian Frugivore." *Proceedings of the Royal Society B* 278 (2011): 3329–35.

Regarding communication between the carnivorous plants Nepenthes and Camponotus ants, see:

—Thornham, D. G., et al. "Setting the Trap: Cleaning Behaviour of *Camponotus schmitzi* Ants Increases Long-Term Capture Efficiency of Their Pitcher Plant Host, *Nepenthes bicalcarata.*" *Functional Ecology* 26 (2012): 11–19.

The Nepenthes rajah also have a close friendship with rats in Borneo, which defecate inside the traps while feeding on the nectar, significantly enriching the plant's diet with nitrogen compounds; see:

—Greenwood, M., et al. "Unique Resource Mutualism between the Giant Bornean Pitcher Plant, *Nepenthes rajah*, and Members of a Small Mammal Community." *PLoS ONE* 6, no. 6 (2011). doi: 10.1371/journal.pone.0021114.

Chapter 5
On the sleep of plants, in addition to Aristotle's previously cited "On Sleep," "On Dreams," and "On Divination in Sleep," see the following:

—D'Ortous de Mairan, J. J. *Observation Botanique*. Paris: Histoire de l'Académie Royale des Sciences, 1729.

—Ray, J. *Historia Plantarum: Species hactenus editas aliasque insuper multas noviter inventas & descriptas complectens*. London: Mariae Clark, 1686–1704.

For further exploration of sleep in *Drosophila melanogaster*, see:

—Shaw, P. J., et al. "Correlates of Sleep and Waking in *Drosophila melanogaster.*" *Science* 287, no. 5459 (2000): 1834–37. www.sciencemag.org/content/287/5459/1834.abstract. doi: 10.1126/science.287.5459.1834.

For an exploration of the abilities of molds to create efficient networks, see this useful article:

> —Tero, A., et al. "Rules for Biologically Inspired Adaptive Network Design." *Science* 327, no. 5964 (2010): 439–42. doi: 10.1126/science.1177894).

Here is the abstract:

Transport networks are ubiquitous in both social and biological systems. Robust network performance involves a complex trade-off involving cost, transport efficiency, and fault tolerance. Biological networks have been honed by many cycles of evolutionary selection pressure and are likely to yield reasonable solutions to such combinatorial optimization problems. Furthermore, they develop without centralized control and may represent a readily scalable solution for growing networks in general. We show that the slime mold *Physarum polycephalum* forms networks with comparable efficiency, fault tolerance, and cost to those of real-world infrastructure networks—in this case, the Tokyo rail system. The core mechanisms needed for adaptive network formation can be captured in a biologically inspired mathematical model that may be useful to guide network construction in other domains.

On the amoeba and its labyrinth-solving ability, see especially the following article:

> —Nakagaki, T., H. Yamada, and Á. Tóth. "Maze-Solving by an Amoeboid Organism." *Nature* 407 (2000): 470. doi: 10.1038/35035159.

On the use of the term intelligence as applied to plants, see:

> —Trewavas, A. "Aspects of Plant Intelligence." *Annals of Botany* 92, no. 1 (2003): 1–20.

As an introduction, here is the abstract:

Intelligence is not a term commonly used when plants are discussed. However, I believe that this is an omission based not on a true assessment of the ability of plants to compute complex aspects of their environment, but solely a reflection of a sessile lifestyle. This article, which is admittedly controversial, attempts to raise many issues that surround this area. To commence use of the term *intelligence* with regard to plant behaviour will lead to a better understanding of the complexity of plant signal transduction and the discrimination and sensitivity with which plants construct images of their environment, and raises critical questions concerning how plants compute responses at the whole-plant level. Approaches to investigating learning and memory in plants will also be considered.

In another article the same author proposes that plants be considered "prototypes of intelligent organisms":

—Trewavas, A. "Plant Intelligence." *Naturwissenschaften* 92 (2005): 401–13. doi: 10.1007/s00114-005-0014-9.

The abstract is quoted below:

Intelligent behavior is a complex adaptive phenomenon that has evolved to enable organisms to deal with variable environmental circumstances. Maximizing fitness requires skill in foraging for necessary resources (food) in competitive circumstances and is probably the activity in which intelligent behavior is most easily seen. Biologists suggest that intelligence encompasses the characteristics of detailed sensory perception, information processing, learning, memory, choice, optimisation of resource sequestration with minimal outlay, self-recognition, and foresight by predictive modeling. All these properties are concerned with a capacity for problem solving in recurrent and novel situations. Here I review the evidence that individual plant species exhibit all of these intelligent behavioral capabilities but do so through phenotypic plasticity, not movement. Furthermore it is in the competitive foraging for resources that most of these intelligent attributes have been detected. Plants should therefore be regarded as prototypical intelligent organisms, a concept that has considerable consequences for investigations of whole-plant communication, computation, and signal transduction.

Also on the theme of plant intelligence, see:

—Calvo Garzón, P., and F. Keijzer. "Plants: Adaptive Behavior, Root-Brains, and Minimal Cognition." *Adaptive Behavior* 19 (2011): 155. doi: 10.1177/1059712311409446.

The article discusses "root brains" and command centers located in the roots, recognizing in plants a certain level of cognitive capacity. The authors write:

Plant intelligence has gone largely unnoticed within the field of animal and human adaptive behavior. In this context, we will introduce current work on plant intelligence as a new set of relevant phenomena that deserves attention and also discuss its potential relevance for the study of adaptive behavior more generally. More specifically, we first give a short overview of adaptive behavior in plants to give some body to the notion of plants as acting creatures. Second, we focus on "plant neurobiology" and introduce the resurfacing of Darwin's idea that plants have a control center for behavior dispersed across the root tips (a root-brain). We then discuss minimal forms of cognition, and consider motility and having a dedicated sensorimotor organization as key features for designating the domain of minimal cognition. We conclude that plants are minimally

cognitive, and close by discussing some of the implications
and challenges that plant intelligence provide for the study
of adaptive behavior and embodied cognitive science more
generally.

On April 10, 1882, Charles Darwin wrote his last known letter, entirely
dedicated to plants, almost as if he meant to seal a life—his own life—
driven by passion for botany. The recipient was James E. Todd, who
was at that time a professor of natural science at Tabor College in Iowa.
Given the letter's brevity, we quote it in its entirety. (Every effort has
been made to maintain the integrity of Darwin's original, handwritten
letter. Abbreviations have been maintained. Italicized and bold words
are used to represent words and phrases originally underlined.)

Dear Sir,
I hope that you will excuse the liberty which as a stranger I take
in begging a favour of you. I have read with unusual interest
your very interesting paper in the American Naturalist on the
structure of the flowers of *Solanum rostratum* & I shd. be grate-
ful if you would send me some seeds in a small box (telling
me whether the plant is an annual, so that I may know when
to sow the seeds), in order that I may have the pleasure of seeing the flowers and experimenting on them. But if you intend
to experiment on them, of course you will not send me the
seeds, as I shd. be very unwilling to interfere in any way with
your work. I shd. also rather like to look at flowers of *Cassia
chamaecrista*.

Many years ago I tried some experiments in a *remotely* analogous case and this year am trying others. I described what
I was doing to Dr. Fritz Müller (Blumenau, Sta. Catharina,
Brazil) and he has told me that he believes that in certain
plants producing 2 sets of anthers of a different colour, the
bees *collect the pollen from one of the sets* **alone**. He wd. therefore be much interested by your paper, if you have a spare

copy that you could send him. I think, but my memory now often fails me, that he has published on this subject in Kosmos.

Hoping that you will excuse me, I remain, Dear Sir

Yours faithfully

Ch. Darwin

P.S. In my little book on the Fertilization of Orchids, you will find under *Mormodes ignea*, an account of a flower laterally asymmetrical and what I think that I called right-handed or left-handed flowers.

To get an idea of the enormous complexity of a single root system, see:
       —Dittmer, H. J. "Quantitative Study of the Roots and Root Hairs of a Winter Rye Plant (*Secale cereale*)." American Journal of Botany 24, no. 7 (1937): 417–20.

For an in-depth discussion of the root tip, see this recent article:
       —Baluska, F., S. Mancuso, D. Volkmann, and P. W. Barlow. "Root Apex Transition Zone: A Signalling-Response Nexus in the Root." *Trends in Plant Science* 15, no. 7 (2010): 402–8.

A recent study of the electrical activity of the root was reported in the following article:
       —Masi, E., et al. "Spatiotemporal Dynamics of the Electrical Network Activity in the Root Apex." *PNAS (Proceedings of the National Academy of the United States of America)* 106, no. 10 (2009): 4048–53.

On the topic of emergent behaviors, hundreds of books have been published, many of them truly fundamental. To explore this fascinating subject, we suggest the following:

—Johnson, S. *Emergence: The Connected Lives of Ants, Brains, Cities, and Software*. New York: Scribner, 2001.

—Wolfram, S. *A New Kind of Science*. Champaign, IL: Wolfram Media, 2002.

—Morowitz, H. J. *The Emergence of Everything: How the World Became Complex*. Oxford: Oxford University Press, 2002.

On the behavior of swarms and the emergent properties of root systems, see the following articles:

—Ciszak, M., et al. "Swarming Behavior in the Plant Roots." *PLoS ONE* 7, no. 1 (2012). doi: 10.1371/ journal. pone.0029759.

—Baluska, F., S. Lev-Yadun, and S. Mancuso. "Swarm Intelligence in Plant Roots." *Trends in Ecology and Evolution* 25 (2010): 682–83.